The OXYGEN REVOLUTION

Geoffrey Mains

David & Charles : Newton Abbot

ISBN 0 7153 5728 X

Set in 10/12pt Plantin
and printed in Great Britain
by W J Holman Limited Dawlish
for David & Charles (Publishers) Limited
South Devon House Newton Abbot Devon

Contents

List of Illustrations

To Manly MacDonald,
with my deepest respect and admiration

PART ONE

Roots

'We shall not cease from exploration
And the end of all our exploring
Will be to arrive where we started
And know the place for the first time.'

T. S. Eliot

CHAPTER 1—1

Coming Home

'. . . for all the night
I heard their thin gnat-voices cry,
Star to faint star across the sky.'
Rupert Brooke

In a single second light can travel nearly 300 million metres; the enormous distance that light travels in a year is termed a light year. So large are the distances of the universe that they are measured in light years, and here on Earth we can detect stars that are billions* of light years away. The matter of the universe is clustered into stars and the stars into galaxies, each moving away from one another at tremendous speeds. Stars range in size from red giants to white dwarfs, and it is in the core of the stars that the density of matter reaches its highest. Indeed in some neutron stars it climbs to trillions of grams per cubic centimetre! Between the stars the density of matter is far less; here and there are clouds of dust, streams of electrically charged particles, and occasionally, about some stars there are planets and moons. Between the galaxies however, there is little.

In the heart of the stars temperatures reach trillions of degrees, in the blackness of space between the galaxies they drop nearly to absolute zero. Between these two extremes are a range of temperatures, for each star is a giant solar furnace emitting steady streams of energy in the form of cosmic, x-ray, ultra-violet, visible, infra-red and radio frequences, all of them travelling at the speed of light, and many of them crossing the blackness of space for billions of years. Stars are born, age and die; some burn out slowly; in contrast others explode violently as supernovae sending shockwaves of energy outwards into the neighbouring reaches of space.

Occasionally, a star like the sun will be circled with planets and the starlight will brush these as it goes by, heating some until they are molten rock, merely touching others with a cold light. But on some

* All use of 'billion' in this book refers to the quantity one thousand million.

planets the amount of solar energy is neither too great nor too little and a complex interaction of this energy with the matter at that planet's surface may occur. These interactions, through billions of years, produce and sustain life. In the universe, life is probably a common process, and one that is likely predictable in terms of physical and chemical laws. Life is a complex system that operates by solar or stellar energy driving matter through intricate but definable pathways. It is a system of incredible complexity and beauty, and a system that through evolution becomes more and more ordered in contrast to the randomness of the universe outside. But most of all life is a conservative and a self-regulating process. In time, in matter and in energy, the universe is an environment of harsh extremes, extremes that are intolerable to earth-bound life, adapted, albeit, by its very biochemical nature to a narrow range of living conditions. These living conditions are rare in space, and indeed probably dependent upon life. Against the great extremes of the universe, life creates and maintains an environment that is conducive to itself. This environment is the space-ship in which life lives: radiation shield, temperature and matter modulator and life-support system—on Earth this is termed the biosphere. Over billions of years of evolution the harsh conditions on Earth have been modified into a narrow range in which life thrives and continues, until the source of energy that spawned it —Sun—fades and dies.

The dramatic story of evolution is the history of our lives, for our very cells, like those of every other living organism today, have their beginnings in a primordial organism that existed about three billion years ago. And this organism itself had a history of another billion years of precarious but essential sub-cellular development. On all of those four billion years we are now dependent, and all have important implications in our lives today. The first part of this book will therefore deal closely with the story of life and with the inter-relationships between all living things, their life-functions, and the quality of their environment.

Only recently has the human animal re-discovered the system of life from which it comes and to which it inevitably goes. The wealth of information gathered by modern science has extended human sight far beyond the confines of spaceship Earth. We have glanced billions of light years to other galaxies and we have looked billions of years backwards through the molecular stirrings of life to the dawn of evolution on Earth. We have glimpsed the core of the atom and the structure of the biosphere. From the contexts of all of these, we have grasped some of the nature of life. Divided from it by culture and technology, we are to a large degree out of sympathy with that world. Yet today we have a technological impact upon the rest of life that threatens our mutual existence. The second part of this book will deal with these impacts and with the fundamental ways in which human thinking and action are at odds with, or fail to take into account, the meanings of our ancestry. To correct this situation, to ensure the survival of life and of human life as an integral part of the biosphere, requires a change of both attitude and action. The third section of this book deals with these changes.

Earth and the Solar System

Over the past decade, modern astronomical techniques and space probes to the planets closest to Earth have told us much of the environments that exist there, and have provided striking contrasts of the conditions that exist in environments where life is non-existent. In their own way these studies have much to tell us of the special nature of the environment on Earth. Against the non-living backdrop of the solar system we shall be able to examine the poem of life.

Table 11—1 provides some details of the nine planets that circle Sun and make up the solar system. They range in size from Mercury, the smallest, to Jupiter, the largest. The farthest from the sun is Pluto, receiving only one-ten-thousandth of the light, per unit surface area, that is received by Mercury, the nearest. Each of these planets requires a different length of time to complete one revolution about the sun, or a year. Mercury travels its orbit in only 88 earth-

Table 11—1 The Planets and Their Dimensions

Name	Mass (in kilograms)	Distance to sun (kilometres)
Mercury	$\cdot 4 \times 10^{24}$	$\cdot 6 \times 10^{8}$*
Venus	$4\cdot 8 \times 10^{24}$	$1\cdot 1 \times 10^{8}$
Earth	$6\cdot 0 \times 10^{24}$	$1\cdot 5 \times 10^{8}$
Mars	$\cdot 6 \times 10^{24}$	$2\cdot 3 \times 10^{8}$
Jupiter	$2208\cdot 0 \times 10^{24}$	$7\cdot 8 \times 10^{8}$
Saturn	$579\cdot 0 \times 10^{24}$	$14\cdot 3 \times 10^{8}$
Uranus	$90\cdot 0 \times 10^{24}$	$28\cdot 8 \times 10^{8}$
Neptune	$93\cdot 0 \times 10^{24}$	$45\cdot 0 \times 10^{8}$
Pluto	$6\cdot 0 \times 10^{24}$	$60\cdot 0 \times 10^{8}$

*10^8 means 100,000,000 or one hundred million

days, while Pluto takes 248 earth-years. At the same time, each of these planets is rotating about its own axis, and the times for this also vary. On Earth we have defined this period as 24 hours or 'one day', but on Venus 225 earth-days are required to complete a single rotation, and on Jupiter only 10 earth-hours.

To a traveller, this system would appear remarkably constant. Cycles of revolution about Sun would continue indefinitely for each planet; so would each planet continue to rotate about its axis and days would pass into nights, each at its own particular speed. As Sun's energy strikes the surface of each planet, different amounts would be received in the same period of time, depending upon speed of rotation, length of day and night, surface area, and the seasonal tilt of the planet towards the sun. The whole picture is one of silent constancy. Even the planets' composition seems to be of the same chemical elements in much the same proportions, the major elements being oxygen, silicon, iron, and aluminium. However, we now know that there are distinctive differences.

Mercury is very hot and apparently has no atmosphere. All types of solar radiation, cosmic, ultra-violet, visible, heat, and radio waves are bombarding its surface, which may still be molten in places. It rotates very slowly, one day on Mercury being equal to eighty-eight on Earth.

Venus, the next planet, shines with a great brilliance. Opaque clouds cloak its surface and reflect back 75 per cent of the sunlight. Its atmosphere is thought to consist mostly of carbon dioxide and hydrocarbons, with traces of mineral acids (hydrochloric and hydroflouric acids) and ice crystals. There is probably also some carbon suboxide (C_3O_2), a whiteish gas.

If Venus were simply to reflect back all of the radiation reaching its surface, a temperature (at the surface) that was just below the freezing point of water would be expected. Instead, the atmosphere at this point reaches a temperature of 500°C, despite the fact that only 25 per cent of the incident solar radiation actually penetrates the atmosphere! This discrepancy has been explained as follows. The atmosphere appears transparent to a fraction of the solar radiation, which thus enters. However, once the radiation has penetrated deep towards the surface, it is reflected back by ice crystals, water droplets, and carbon suboxide, and transformed into infra-red or heat radiation. Since infra-red radiation is readily absorbed by carbon dioxide, the atmosphere now appears opaque to the reflected sunlight, which is therefore detained. In this way it is bounced back and forth within the atmosphere, being lost to space only at a very slow rate. In addition the surface of Venus may not have solidified completely, and the heat given off from cooling is also trapped. The atmosphere of Venus could thus be likened to a greenhouse (see Figure 11—1).

Beyond Earth, the next planet outwards from Sun is Mars. Here the atmosphere is almost totally carbon dioxide, and there are no

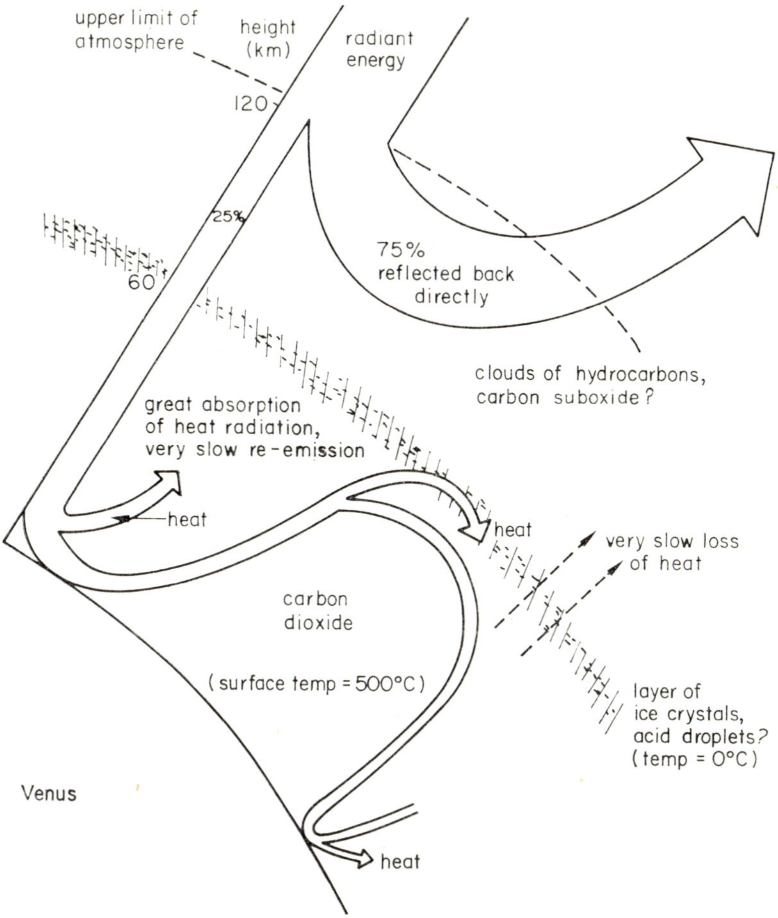

Figure 11—1 The Greenhouse on Venus

detectable traces of nitrogen or water vapour. Recent space probes have shown that the surface temperature of Mars is about 50°C below zero. This provides a good explanation for the fluctuating polar caps on Mars, which have been traditionally viewed as ice caps like the ones on Earth—on Mars the temperature occasionally drops to at least 80°C below zero and carbon dioxide precipitates out of the atmosphere as dry ice. The polar caps advance and retreat with the seasons on Mars. Recent aerial surveys of Mars conducted by spacecraft have confirmed this explanation.

Mars, unlike Venus, has very little opacity to its atmosphere. Thus, although both atmospheres largely consist of carbon dioxide which strongly absorbs and detains infra-red radiation, on Mars the greenhouse effects are much reduced and go only part of the way towards overcoming the great distances between Mars and Sun. The surface of Mars is likely bombarded with many types of solar radiation, including the powerful ultra-violet frequencies.

Jupiter, the giant of the solar system, lies at a distance five times that of Earth from Sun. The atmosphere of Jupiter probably consists of ammonia (NH_3), methane (CH_4), hydrogen (H_2), and helium (He). Gravity on Jupiter is very high and the core of this planet is thought to consist of a condensed mixture of metallic elements and hydrogen, but in a form vastly different from the metallic rocks of the other planets. The density of Jupiter is extremely low, being only 1·5 times that of water! The surface of Jupiter is thought to be covered with layers of frozen water and ammonia. The planets beyond Jupiter—Saturn, Uranus, Neptune, and Pluto—are probably cold and desolate. They receive meagre rations of sunlight, and some of them are surrounded by many moons; however, beyond this we know very little about them.

Earth appears strikingly different from the other planets. Its surface is partially covered with a colourless liquid, water, which appears blue from the reflection of sunlight; and its atmosphere is filled with constantly moving patterns of water vapour or clouds. Earth's distance from Sun means that on the basis of incident sunlight alone, the surface temperature would approach 45°C (113°F). Earth has an atmosphere unlike the others in the solar system, an atmosphere that contains nitrogen, carbon dioxide, and a chemically reactive material—free oxygen. It also contains large amounts of water vapour, which vary with altitude, temperature, and surface position.

Table 11—2 The Composition of Earth's Atmosphere

Component	Chemical formula	% by volume
Oxygen	0_2	20·95
Nitrogen	N_2	78·09
Argon	A	·93
Carbon Dioxide	CO_2	·03
Neon, Helium, Krypton	Ne,He,Kr	traces

As well, there are minute traces of other materials such as hydrogen sulfide, sulfur dioxide, methane, nitrogen oxides, and ozone.

From surface to space, Earth's atmosphere is divided into a number of distinctive layers. From the surface to a height of about twelve kilometres exists the troposphere, the zone in which most cloud formations occur and weather takes place. From twelve to about thirty kilometres is the stratosphere. Most of the gases in the atmosphere exist below the stratopause—the upper limit of the stratosphere. Above this layer, the inward pulling forces of Earth's gravity become weak and the atmosphere becomes much thinner. Above the stratopause the mesosphere extends to about a hundred kilometres, and above the mesopause, for another three hundred kilometres, lies the ionosphere. As its name implies, the ionosphere consists largely of ions, or charged matter. The ionosphere is the last region of the atmosphere before space. It receives full force, the direct radiation of Sun. It is because of these high-energy conditions that chemical compounds break up to form charged particles. And because the most potent forms of radiation—x-rays, gamma rays, and cosmic rays—are absorbed at this point, they fortunately do not reach the Earth's surface. The ionosphere is really the first of a series of windows that comprise Earth's atmosphere, as illustrated in Figure 11—2.

As well as filtering out the more energetic forms of radiation, the ionosphere filters out some of the radio waves in the sun's energy. Visible, infra-red, and ultra-violet radiation chiefly penetrate this layer. Then the energy stream hits the mesosphere. At this point the powerful ultra-violet radiation hits oxygen molecules, producing ozone as a product. As a result this radiation is dispersed here and a layer of ozone is maintained. Then, at the upper troposphere, clouds

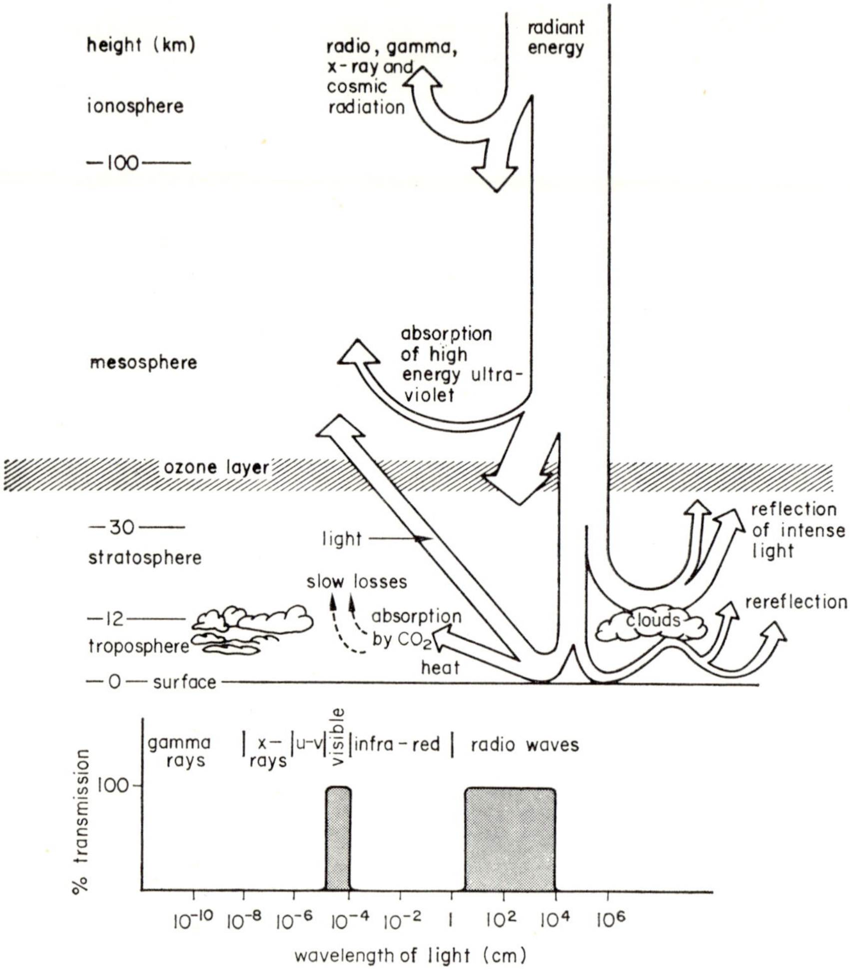

Figure 11—2 The Atmospheric Window

reflect back visible light, particularly that of high intensity. Some is absorbed, causing patterns of motion among the clouds, which further disperse the energy. Heavy sunlight beating down on the waters of the planet induces cloud formation and in the process the solar energy is dispersed indirectly through the motion of clouds, at least partially protecting Earth's surface from a continuous blast of direct sunlight.

Together all of these mechanisms produce an atmospheric window, which is illustrated at the bottom of Figure 11—2. Of a vast range of electromagnetic radiations emitted by Sun, only a small portion reaches Earth's surface. Except for a large section of the radio-wave band, most of this consists of visible light.

The radiation that reaches the surface of Earth has two possible fates. When light hits an object, depending upon the chemical nature of the object and the character of the radiation, some of it may be absorbed, while the remainder is simply reflected outwards, giving the object its characteristic colour. Eventually the fraction of light that has been absorbed is itself radiated out, but this time as infra-red radiation. A bright sandy beach will reflect back most of the light that strikes it, whereas a dark soil will absorb most of it. The heat energy that is radiated back into the atmosphere can be absorbed by carbon dioxide in the atmosphere before escaping the planet. This process has a warming-up effect. Light reflected from Earth's surface can often be reflected back again from the underside of clouds, some of it as heat. Here another portion of the visible radiation is becoming heat radiation and helping to warm up the planet's surface.

As the temperature of the atmosphere rises, the amount of water vapour increases in response. Thus the opacity of the atmosphere increases and this too helps with the retention of heat. An intricate balance between carbon dioxide and water-vapour levels in the atmosphere is therefore effective in maintaining the heat balance of Earth. Perhaps part of the reason that Earth does not become a greenhouse like Venus is the important dispersal mechanisms of air and water currents. Their motions help to distribute heat more evenly over the surface of the planet, so that even if Sun is directly overhead, temperatures will remain at a moderate level.

Beyond this, the atmosphere, the oceans, and to a certain extent the land masses have a heat-retaining ability and thus only slowly lose their heat during the night.

All of these factors interact together to keep world surface temperatures, for the most part, in a $-30°$ to 40°C range. Exceptions, when they occur, are usually localized ones. The importance of this is immense, because at the atmospheric pressures found near to Earth's surface, water boils at 100°C and freezes at 0°C. It is thus possible for water to exist on Earth in all three states—gas, liquid, and solid. As a result, weather is possible and most of Earth's surface is continually irrigated. Erosion can thus occur, producing fine particles of rock which are available for soil building. At the same time, water is made available to areas even at the centre of land masses.

In comparison with the other planets in the solar system, the conditions on Earth are mild.

Let us turn back five billion years to the time when the solar system and Sun were formed. As one current theory goes, Sun and the nine planets, their moons, and the asteroids condensed from a cloud of interstellar dust. Each of the hot masses was drawn towards Sun by the great forces of its gravity. At the same time each mass had a tremendous forward force of motion in space, at right angles to the pull towards Sun. Together these two forces set each planet in a specific trajectory, and unhindered they have spun thus ever since.

Each of these planets slowly cooled, as Sun's radiation evened out to a steadier flux. The planets at the greatest distance probably cooled the fastest. As each planet cooled, it shrank, and a crust of rock appeared on its surface. About each planet there were probably primary atmospheres of helium and hydrogen, but these being very light were soon lost to space. Exuded from the cooling crusts the secondary atmosphere developed, composed of carbon dioxide, ammonia, methane, hydrogen sulfide, hydrogen chloride, and water vapour. There was no free oxygen: at the high temperatures that must have existed, oxygen, being a very chemically reactive element, would have combined almost completely with other elements, for example with hydrogen to form water, with silicon to form silicates (sand), and with carbon to form carbon dioxide.

From that point in time until now, the matter on each of these planets has slowly become chemically stabilized through reorganization, to form the present conditions. Mercury probably lost most of its atmosphere to Sun's gravity. Venus may have lost the ammonia and some of the water, while the remaining water formed ice crystals. Mars became almost totally cloaked in carbon dioxide. But on Earth, something quite different happened.

CHAPTER 1—2

Our Primordial Beginnings

I have heard the summer dust crying to be born
As much as ever flesh cried to be quiet.
Robinson Jeffers

From modern evidence it is apparent that the changing force of Earth's environment was life. The oldest fossils of higher life-forms, like simple seaweeds and jellyfish, date only from half a billion years ago. Fossilized bacteria and one-celled plants have been found that existed before this time, and the oldest bacterial remains yet known are those in the Fig Tree Formation of South Africa. The first stirrings of life probably stretch back to about four billion years ago. We can only imagine these beginnings and the ways in which the first simple cells appeared and developed. Little can be stated as definite fact. Our ideas are drawn together from current theories based upon evidence gleaned in disciplines as far apart as biochemistry and geology. Yet from our present knowledge of life and the history of our planet, there is a good possibility that many of the events that will be described here happened in these or in similar ways. From the very beginning of life it appears that the processes by which living organisms functioned were inexorably bound to the changes that occurred in the environment.

Four billion years ago the atmosphere of Earth was a toxic mixture of gases: methane, carbon dioxide, ammonia, hydrogen sulfide, hydrogen chloride, water vapour, and possibly some hydrogen and nitrogen. The atmosphere and the surface of Earth were bombarded with intense radiant energy from a newly formed Sun, including intense ultra-violet, infra-red, and visible wave-lengths. Indeed, these radiant conditions were probably so intense that most of the living organisms of today would have had to live ten metres (thirty feet) underwater in order to escape their death-dealing effects.

As the surface of planet Earth cooled, it shrank. Fissures formed in the crust from which molten rock burst over the surface of the planet. As cooling occurred, massive amounts of steam were exuded from the crust and collected in the atmosphere. Intense electrical discharges occurred between the atmosphere and the surface of Earth.

As cooling progressed, some of the steam began to condense and eventually great rainfalls took place; the water only to be evaporated again by the hot rock. Gradually however, enough cooling had occurred that water began to collect in small pools, which grew steadily with the continuing rains. The waters dribbled over the rock, cutting pathways in it and beginning the process of erosion; the pools in which these waters collected became rich with dissolved minerals.

About this time, other events began to take place. The conditions

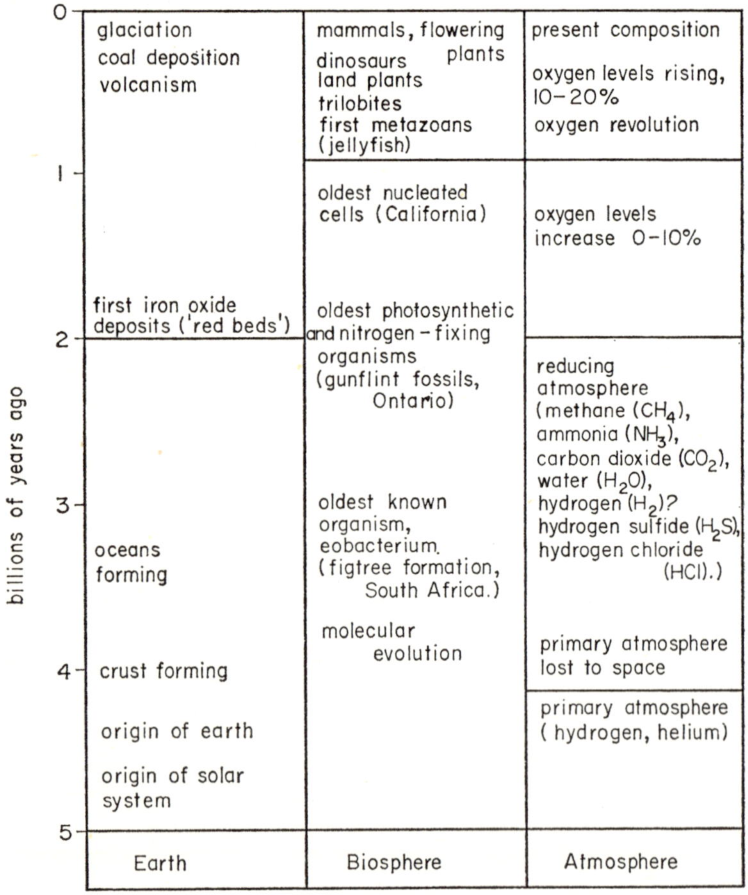

Figure 12—1 The History of Life

of intense energy—lightning, thunder, ultra-violet, visible, heat, and cosmic radiation—along with the presence of newly formed mineral surfaces all began to catalyze chemical reactions among the various components of the early atmosphere. The ammonia, methane, carbon dioxide, hydrogen, and water all began to react in different ways to form new molecules. These molecules were probably broken up by the intensely energetic conditions that led to their formation. But they re-formed, and the cycle of formation, breakdown, and re-formation had begun. As the earth continued to cool and water collected, many of these compounds began to last a little longer and to collect in solution.

What were these new molecules? Today they still appear on planet Earth, as the basic subunits of all life from the measles virus to the antelope. They are, for example, the twenty different types of amino acid molecules, which strung together make up proteins and enzymes. Or they are the nucleotides, which strung together constitute the nucleic acids, the molecules of heredity. Or they are the sugars, which strung together make up cellulose (wood) or starch. Or they are the pyrrole rings, which strung together in fours form the oxygen-carrying heme group of blood cells and the photosynthetic pigment chlorophyll of green plants. Today all of life employs the same subunits, because all of life came from these chemical changes in the environment.

As these molecules began to form and to collect, they also began to interact with each other. In pools of warm water, they became linked together to form larger molecules, or polymers. These were akin to the macromolecules like cellulose and proteins of today. Because of their properties, these polymers had a greater affinity for one another than for the aqueous medium they were dissolved in; they therefore began to aggregate in droplet-like formations. Many of these droplets had some of the simple properties of cells. They grew and shrank with warming and cooling; they preferentially collected certain types of atoms and molecules in their interiors while excluding others. Some of them even began to split up and to collect new material, as well as losing some. The cycle went on, form, break up, form.

At this point some of the droplets (formed of aggregates of proteins, fats and carbohydrates) began to acquire catalytic properties. That is, they began to be able to carry out certain chemical reactions themselves, and for some of them this was exceedingly beneficial. By carrying out such reactions they could build their own starting materials from those in the environment, without being dependent solely upon chance. This then was the beginning of life-forms as we know them today. The process leading to this form of life was not a fast one. Over a period of about a billion years there were many false starts. Components formed in the environment, aggregated to form polymers and droplets, even gained catalytic powers, yet survived no more than a few days beyond this. The process was probably repeated millions of times, through many different variations of the basic environmental material. Eventually more viable units were bound to appear. But of all these, only one proceeded onwards, to give birth to the rest of life.

The catalytic properties acquired by the ancestral organism were essential to its survival. And these properties were determined by the chemical natures of the molecules making up this organism—proteins, lipids, nucleic acids, and carbohydrates. They remained essential for a very long time—in fact many of these molecules were conserved during the billions of years of evolution that followed in which this 'proto-organism' gave rise to the many diverse organisms of today. Looking at the molecular processes by which life functions today, one is greatly surprised: despite the great variation in shapes and sizes, biological roles, and feeding habits, all organisms work on variants of a basic theme. Glucose, for example, an energy-providing sugar, is metabolized by the same eleven steps, involving eleven homologous enzymes, in a rat cell and in a yeast cell. There are variations from organism to organism, and these are the reasons why life has the diversity it does today. But these variations occur about a basic theme. The deeper that one delves into the molecular structure of a cell, the more obvious the similarities become. In the nucleus or central region of a cell is the information required to construct that cell, and if the cell is part of a larger organism, also the information for the whole organism as well. The blueprint for the organism is stored in structures called 'chromosomes', and the message is coded in a language containing four letters and sixty-four three-letter codons. All living organisms use the same four letters and understand each of the sixty-four codons in the same manner. It is even possible for the machinery in a mouse cell to read the messages written on a human chromosome, and to translate these messages into protein matter that otherwise would have been found only in a human cell. (See Figures 12—2 and 12—3.)

Some of the earliest mechanisms of life were probably geared to obtaining energy in a usable form for growth and reproduction. The method that came to be used involved the splitting of the carbon-

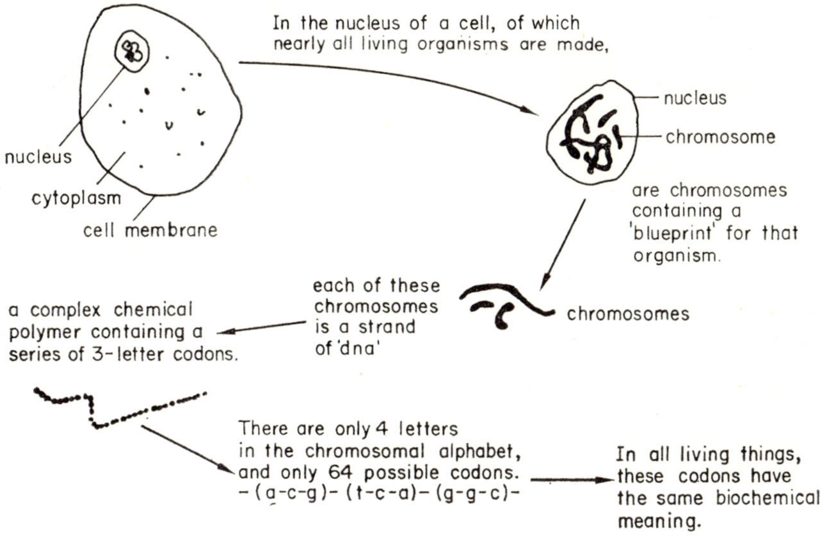

Figure 12—2 The Genetic Code

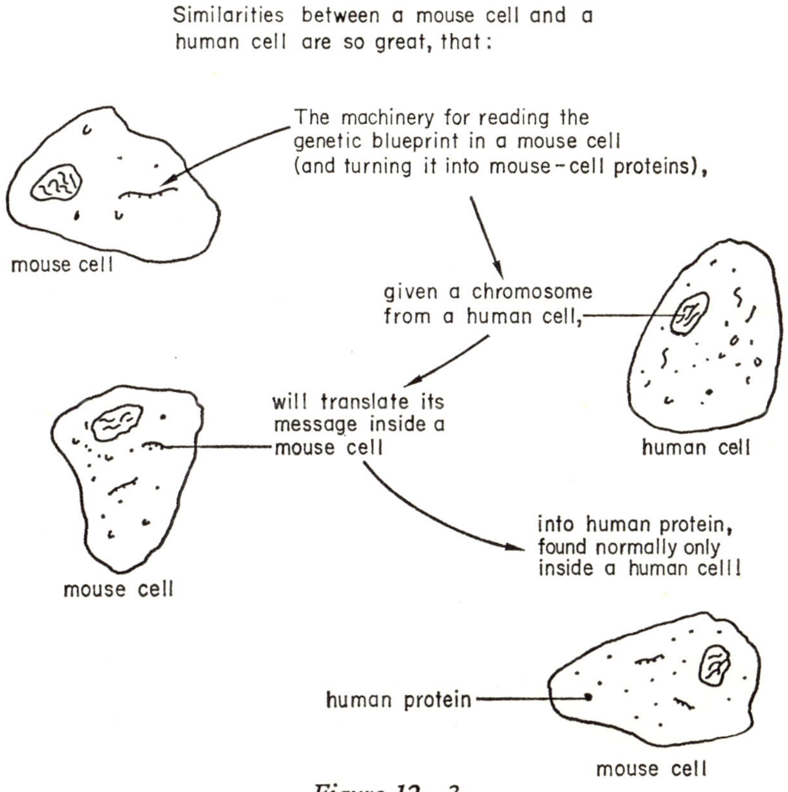

Figure 12—3

carbon bonds in life itself and the transferring of the energy that was released to a chemically desired bond. The process of fermentation (carried out by yeast cells where there is little or no oxygen) may have been employed. So also may have been a process where dissolved sulfate ions (SO_4^{2-}) were converted to sulfides (S^{2-}) while the carbon-carbon bonds went to carbon dioxide. This process, however, was very limited in scope, because the availability of sulfate was limited. Yet today these processes are still employed by many micro-organisms.

The largest step forward in energy-gathering was the evolution of

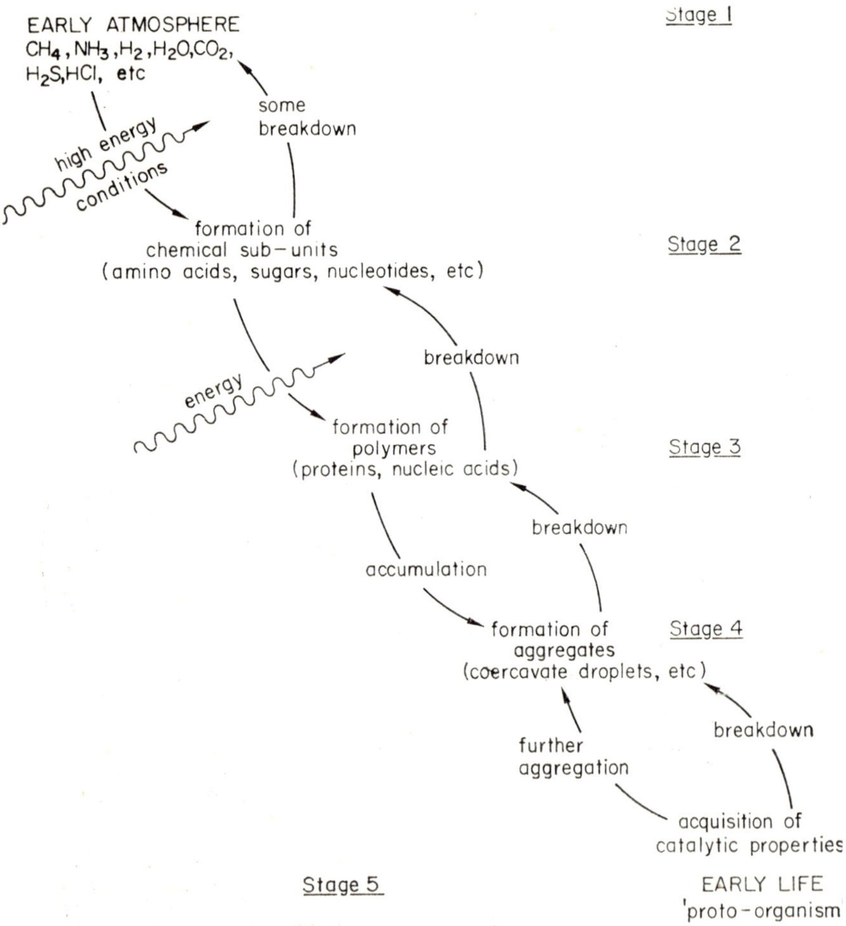

Figure 12—4 First Steps in the Origin of Life

photosynthesis, attained by the ancestors of modern green plants. Through this process, the production of food-material was continually possible:

$$\underset{\text{carbon dioxide}}{6CO_2} + \underset{\text{water}}{12H_2O} \xrightarrow[\text{Enzyme Systems}]{\text{Light Energy}} \underset{\text{glucose}}{C_6H_{12}O_6} + \underset{\text{oxygen}}{6O_2} + 6H_2O \quad (2)$$

In other words, carbon dioxide and water are converted by light and by enzymes into glucose and oxygen.

All of the basic components of this reaction were common in the environment, and the catalytic power necessary to carry it out is found in the complex enzyme systems of green plant cells. Chlorophyll is an essential molecule in this process—and the reason plant cells are green. The chlorophyll molecule is important in the trapping of sunlight energy in the visible range. This energy is consequently transformed into chemical bonds between carbon dioxide and water molecules. The products are a sugar called glucose, and free oxygen.

To use glucose, in other words to get at the energy trapped in the bonds, an opposite process termed 'respiration' evolved along with photosynthesis. It appears that the primitive methods of obtaining energy from the environment by using sulfate and other materials may have been an early form of respiration. In time this process gave rise to photosynthesis and the form of respiration that complemented it:

$$C_6H_{12}O_6 + 6O_2 \xrightarrow[\text{systems}]{\text{cell enzyme}} 6CO_2 + 6H_2O + \text{energy for cellular use} \quad (3)$$

Respiration thus uses the oxygen produced by photosynthesis to produce the starting materials—water and carbon dioxide. Again, this process is carried out by enzyme systems in living cells. The enzymes of respiration are located in subcellular structures called 'mitochondria', similar to capsules termed 'chloroplasts', which contain the machinery of photosynthesis. Since many of the same enzymes exist in both the chloroplast and the mitochondria, it has been suggested that both of these systems have a common ancestor. Current evidence even suggests that this ancestor was once a free-swimming micro-organism itself, that with time became incorporated inside the cell walls of other unicellular organisms.

A new type of organism began to appear in the environment about this time. This organism lived solely off the live or dead photosynthesizing organisms, thus depending upon the green plants for its food. There were probably three main types of life at this point. First, there were the earliest forms, living off processes such as fermentation and sulphate reduction for their energy—these were the forerunners of today's bacteria. Secondly, there were the photosynthetic organisms, able to produce their own plants. Thirdly, there were the first herbivores—the green-plant feeders—ancestors of today's animal kingdom.

The processes of photosynthesis and respiration had distinct advantages to primitive life. No longer was there any reliance, as with the earlier organisms, upon the slow and tedious process of environmental catalysis, the only means by which the basic subunits of life could be gleaned. Organisms were now able to make their own food and their own building-blocks. Thus these photosynthetic organisms must have rapidly proliferated and an excess of free oxygen was produced.

Life in the warm seas continued to flourish and organisms were born, grew, reproduced, and died. Their dead remains succumbed to one of a number of possible fates. Some were consumed by the primitive fungal and bacterial organisms that broke down their complex structures and used the chemical bonds in them as sources of energy. The oxygen released initially during the production of these bonds was therefore reconsumed during this process of respiration, and carbon dioxide was released. Others of the dead organisms sank to the sea floors where they were buried in the mud and ooze beneath the slowly accumulating sediments from erosion of the earth's surface. Here they were trapped outside the biological world, and the carbon-carbon bonds in them—formed by photosynthesis—were not broken down again. Thus the oxygen produced during the initial photosynthesis of these bonds was not reconsumed and represented a net gain to the environment.

As the first oxygen was produced, it began to seep out into the surrounding environment. At first it remained dissolved in the sea water occupied by the first photosynthesizing cells, but as the water became saturated oxygen began to bubble out and came into contact with the atmosphere. Here a wide variety of reactions was possible because of the high reactivity of molecular oxygen, and with its slow production over the next billion years or so, great changes began to occur. Methane and ammonia began to disappear from the atmosphere as they reacted with oxygen to produce carbon dioxide and nitrogen oxides. The latter were washed out of the atmosphere by rain, producing nitrites and nitrates that began to collect in the soils and the oceans.

Certain living organisms evolved enzyme systems that could take advantage of the free oxygen in the atmosphere. Certain bacteria used molecular oxygen to convert ammonia to nitrogen, nitrites, or nitrates, while gleaning the energy released during this process. Other bacteria evolved that converted sufides to sulfur, sulfites, thiosulfates, and sulfates, again with the use of free oxygen. Still others may have converted hypophosphites and phosphites to phosphates.

With all of these processes, the raw material of the environment was being converted to new forms, and for many of the very primitive organisms, still dependent upon the early environmental conditions, this was a disaster. Other primitive organisms, however, could benefit. The bacteria that used oxygen to convert sulfides to sulfates were actually producing a foodstuff for the bacteria that in the very first processes of fermentation had converted sulfates to sulfides in the absence of oxygen. Similarly, primitive microbes existed that could convert nitrates to nitrogen in the absence of oxygen, and these organisms likewise benefited.

However, as the methane and ammonia in the atmosphere began to dwindle, even these new organisms, the users of free oxygen, became dependent upon the wastes of the photosynthesizers and respirers. For with the changing nature of the atmosphere these wastes became the only sufficient sources of sulfide and ammonia. And also associated with these wastes, but existing where conditions were lowest in free oxygen and most like those of the pre-oxygen atmosphere, were the remaining primitive organisms. Thus, all living organisms —the primitive bacteria, the oxygen-using bacteria, the photosynthesizers, and the respirers—became tied together in an interlocking cycle of life by the very nature of their needs for various nutrients and their life-function. Even today many of these bacterial organisms play much the same roles in the conversion of nitrogen and sulfur atoms from one form to another, that they learned two billion years ago.

Between these primitive organisms and the environment, flows of matter were beginning to develop. Nitrogen atoms, for example, followed certain pathways. (See Figure 12—5.) Ammonia in the atmosphere (1) was converted by chemical reactions to the constituents of life. Early plants and bacteria (2) absorbed these subunits. Animals eating these plants then assimilated the fixed ammonia (3). As these

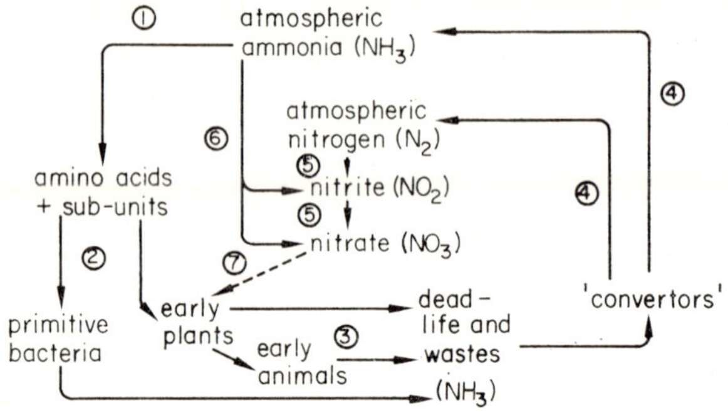

Figure 12—5 Early Nitrogen Cycling

plants and animals died, their dead bodies provided food for the converters and fermenters, who either released the nitrogen as ammonia or converted it to molecular nitrogen (4). The converters, with the use of atmospheric oxygen, changed free nitrogen further to nitrites and nitrates (5). Some of the green plants adapted to the use of nitrates as a source (7). Atmospheric oxygen was converting some of the ammonia to nitrites and nitrates (6). Through time, as the concentration of atmospheric ammonia dropped, the living section of the cycle became more and more important as an influence in the biosphere. Molecular nitrogen (N_2) became the major component of the atmosphere, and, except for that in living systems, nitrite or nitrate became the only other major forms of nitrogen.

Carbon, too, followed certain pathways (Figure 12—6). In the initial stages methane and carbon dioxide went into life constituents (1), and these were absorbed (2) by the early bacteria and plants. Photosynthetic plants on the other hand used atmospheric carbon dioxide (dissolved in sea water) as a food source (3). These were eaten by the early animals (4), and the death of both of these provided matter for bacterial organisms. These organisms—plants, animals, and bacteria—turned the fixed carbon into carbon dioxide through respiration (6), using either sulfate, nitrate or oxygen. Methane was slowly converted in the presence of free oxygen to carbon dioxide (7). Some of the carbon fixed in living organisms was lost to the deep ocean sediments (8).

The atmospheric concentration of methane thus decreased and the concentration of carbon dioxide rose. But being very soluble in water, the carbon dioxide remained to a large extent dissolved, and some

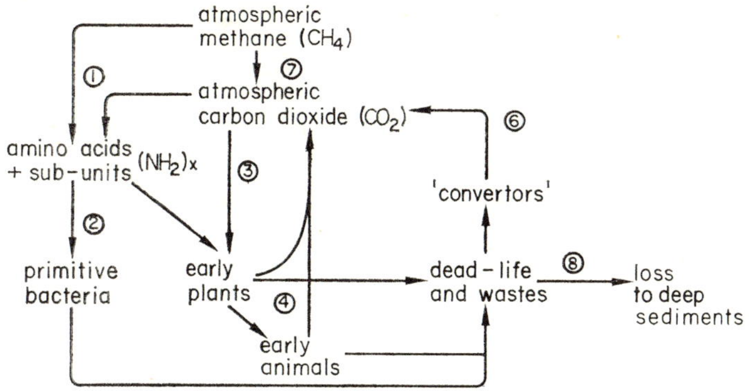

Figure 12—6 Early Carbon Cycling

was removed by precipitation as calcium carbonate to form limestone. This process occurred in slightly alkaline waters and was accelerated by organisms such as coral, which use limestone to build their structural supports.

All of these changes in the flow of matter did not occur quickly, but over three and a half billion years. This is a long time period—seven times the period that was necessary for the evolution of whales and oak trees and other complex organisms from simple plants and animals. This long period was necessary because the life that was forming and carrying on during it was doing so against great odds. The environment was hostile and bombarded with heavy radiation. The slow process of organization on a subcellular level, much of it reliant upon chance, had to be complete before any of the more complex forms of life that we know today could be possible. And the incident ultra-violet radiation was highly destructive to the complex and ordered molecules that make life possible.

Through this long period of time stabilization was beginning to take place in the environment—and the forces behind these changes were the matter flow systems set up by life itself. Many of the materials—ammonia, methane, hydrogen sulfide, and phosphites—that are toxic to life as we now know it were being reduced to trace amounts. More important, over a period of just greater than a billion years, the oxygen revolution was occurring. As free oxygen built up in the atmosphere, an ozone layer began to form that moved upwards through the atmosphere with the increasing concentration of oxygen, until it reached the mesosphere. Earth's surface was thus shielded from ultra-violet radiations, and the upper regions of the oceans and the land masses became habitable. With this transformation came a great blossoming of evolution into the millions of diverse forms of life that inhabit Earth today.

CHAPTER 1—3

Movement at the Still Point

The last half-billion years of evolution were perhaps the most spectacular of all. They are certainly far better known today than the earlier period of subcellular organization. From the ages of the trilobite, the carboniferous swamps and coal forests and the dinosaurs, abundant fossil remains tell of the changes leading to present-day life-forms. Although the most striking of these changes seem to have taken place largely at the level of the organisms themselves, whether as features of body coloration, or nervous function, or anatomy, these changes in every case reflected the nature of the genetic messages that are stored in the heart of each cell in an organism. It was upon the themes contained in these messages that evolution played its many variations.

Reading the Blueprint of Cells

There are contained in the cells of every organism from the unicellular amoeba to the multicellular wheat plant a group of large molecules that hold blueprints for all the workings of that organism. (In many cells, these molecules are held in a well-defined region termed the *nucleus*; in bacteria, distinctive nuclear regions are not present, and in some highly differentiated or specialized cells such as the human red corpuscle, they have disappeared.) These blueprints, as we have seen, are in the form of DNA molecules—long polymers of four different chemical compounds strung together in a spiral. The order in which these beads are threaded is important, for it is this order that spells the message that the DNA contain. There are four different letters in this spiral bead-alphabet, and grouped in threes these beads spell out the assembly instructions for specific proteins. The DNA molecules containing the cellular blueprint are arranged in structures called *chromosomes*, which are readily visible under a light microscope. Each type of protein that an organism is capable of assembling seems to be represented on the chromosomes by an area called a *gene*. The proteins, either structural components of the organism, like hair or muscle fibre, or enzymatic components like digestive enzymes, are thus involved in controlling the functional activities of the cell and the organism. Some of these proteins may determine macroscopic and easily visible features like the colour of hair or of eyes, or the blood-group type. In this way, the features are said to be controlled by genes. Other features however, like facial shape, are the result of many genes working together. On the other hand, many genes are not reflected in macroscopic features, but as biochemical features within one type of cell, for example, liver cells.

Many genes may simply not be expressed in an organism, they may remain muted for varying periods of time. Some, for example those involved in mammals during the development of resistance to disease, may only be expressed when a particular bacterium has been present and has induced the gene to action. Others may be operative during only a certain period of life, for example during seed germination, during infancy, or during flowering. Still other genes may never be expressed. There appear to be two great forces at work in any organism, and these are genetic and environmental. The genes of an organism determine the range of possible patterns that an organism can take; environmental factors during growth and adulthood then mould these patterns into the individual. They cannot, however, stretch the organism to degrees beyond that stated in the blueprint.

The physical nature of every organism, then, is merely a reflection of the genetic messages carried in the nuclei of its cells. When the cell is ready to divide, its first job is to reproduce the nuclear messages so that each of its daughter cells will contain a complete set of instructions for cell maintenance. After this is done, the cell can divide, and in the process of mitosis the nuclear material—contained in the chromosomes—is pulled equally to each side of the split.

However, during the process of replicating the nuclear library, changes can occur in the message. One of the words may not be put in the correct order, or a word may be omitted, added, or altered. Many of these mistakes appear simply to be chance slip-ups in the workings of the cell. Others can be induced by high-energy radiation or by carcinogenic materials like benzopyrenes. All of these changes are termed *mutations*.

Because of mutation, a daughter cell may have a set of messages that differs slightly from those of the parental cell. There are a number of possible consequences of mutation. Should it, for example, occur in a vital part of the message, the daughter cell will not survive.

Say, for example, that we are looking at a part of the nuclear library that codes for an enzyme essential for the cell to make use of glucose. If a mutation in this part of the library changes the message so that this enzyme product comes out inactive, then the organism will not be able to consume glucose and will therefore die. However, suppose that the mutation occurs in another part of the library, on a message for a defensive protein, one that is vital in the case of an attacking bacterial organism. Only if that bacterium attacks will the mutation be lethal, otherwise it will go unfelt and can be transmitted from generation to generation. Some mutations may actually be harmless, while others may be beneficial.

It should be noted that mutation occurs in cells in the resting state, as well as in dividing cells. The rate of mutation is usually lower, because to a certain extent the chromosomes are less exposed. Here too, mutation may be caused by ionizing radiation or materials like benzopyrene.

For a long time genetic studies have indicated that a certain amount of background mutation is always occurring. At every cell division some changes probably occur. And recent biochemical studies of the actual messages contained in the nuclear material tend to confirm this view. The rate of mutation however is not a constant phenomenon. It appears to be higher in certain environments than in others. (Cancers appear to be characterized by abnormal rates of cell division, caused in some cases by high rates of mutation—for example from the presence of carcinogens in automobile exhaust and tobacco smoke.) Mutation through geological time has probably varied considerably.

In the early years of Earth, before the presence of an oxygen-rich atmosphere, large doses of ultra-violet radiation very likely accelerated the mutation rate. Today, mechanisms in the nuclei of cells still exist to repair damage caused to DNA molecules by ultra-violet radiation. Some of these mechanisms may be hang-overs from this early time. Through continual and random background mutation, the *genome* (the total genetic structure) of a population of organisms shifts from one of homogeneity to one with a broad variety of individuals, each differing in slightly different ways. This genetic heterogeneity—which is constantly changing shape with background mutation—has made evolution possible.

The Forces of Selection

Let us return in time to the point at which the first simple photosynthetic organisms such as the phytoplankton appeared. Let us look at a population of these organisms living along the warm shallow coastline of a continent. The temperature of the water in which these phytoplankton live is very important, for they tend to function optimally at 30°C. Figure 13—1 illustrates this. In a population of these

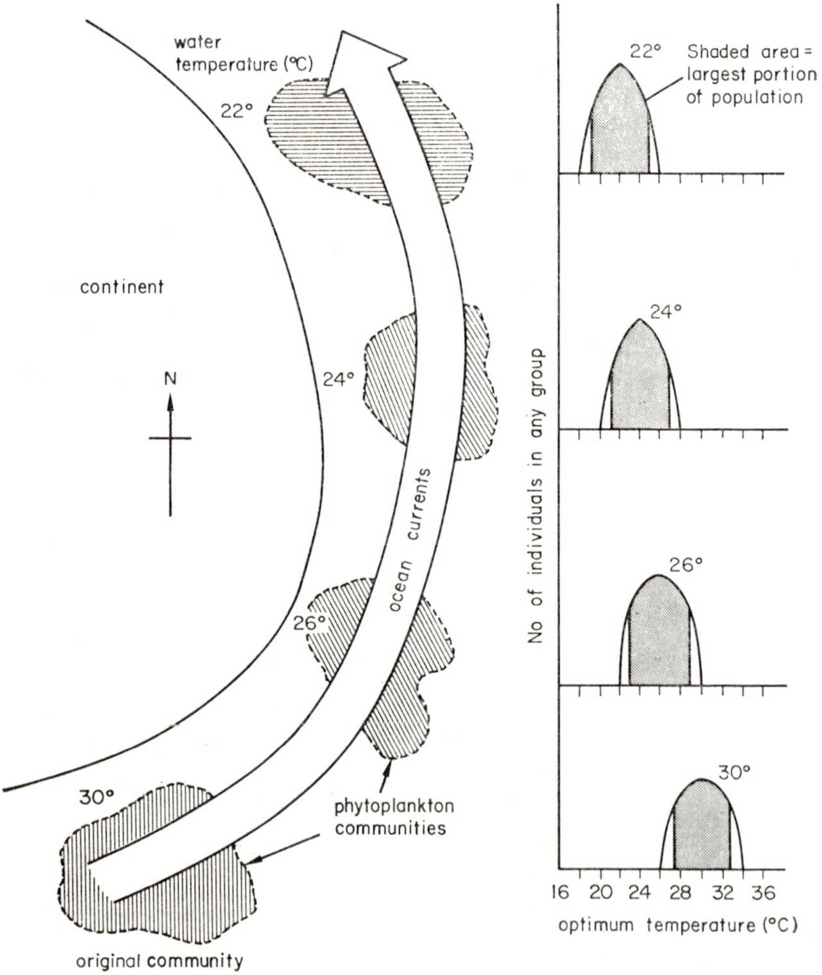

Figure 13—1 Forces of Selection

phytoplankton, most function best over a range of 28 to 32°C. But because of background mutation, the population contains a few individuals who function optimally outside of this range. These organisms will continue to be minor components of the population in the warm shallow seas, because they are functioning here under what are to them unfavourable conditions.

Along this coastline there are moderate currents moving in a northerly direction, carrying the warm waters of the south into progressively colder waters as they move northward over a thousand-mile strip. Between the cold northern and warm southern waters then, there exists a temperature gradient. As time progresses, some of the phytoplankton of the southern waters are carried northward by the currents. As they progress northward through the gradually decreasing temperatures, selection occurs. The populations in each area are formed mostly of the organisms best suited to the ambient temperature. By the time the phytoplankton have reached the northern waters, the population has a completely different temperature optimum from that of the southern waters. It is now 22°C and none of the phytoplankton at the original optimum (30°C) have survived. This slow change in the genetic composition of the phytoplankton population occurs over many generations and shows the way environmental forces select certain traits or features from the genetic pool.

Selection can occur in many ways. The selecting force may be climatic or geological, or it may be the presence of another biological organism. Another species of phytoplankton for example, may appear in the warm shallow seas. It may outcompete the original variety in the waters near the surface, selecting from it a strain that functions best at a depth of two to three feet below the surface. At any one time, selection may be occurring on a wide number of different parts of the genome. Selection at any one of these points may be occurring under the influence of one or more forces generated by the biological or physical environment. Selective pressures act then upon the genomes of any population by selecting from among the many combinations of background and random mutations those individuals and strains that are best suited to survival under the changing conditions. Again, selection is not a short-term process. It does not produce dramatic changes in a single individual, but works over periods of many generations. In the long term, over a thousand-year period, it may lead to the origin of new species.

A species is considered to be a population of organisms whose genomes are so similar that it is only possible for individuals within the species to mate successfully. *Speciation*, the phenomenon that produces species, is a process that brings about this state of reproductive and genetic segregation.

The process of speciation may occur by a number of means (one of which is selection). Physical isolation is an important mechanism in this process. Should the seeds of a certain plant be carried by water or wind to a point many miles from the original community, then a new community may arise which will grow in its own directions, changing under the selective pressures that exist in the new location and that may be very different from the pressures in the home community. As time goes on each community will develop characteristics that are distinctive to it alone, and over a very long time distinctive species may emerge. At this point genetic differences between the two communities will have become so great that cross-fertilization is no longer possible. Two species have thus emerged from one.

On the other hand, speciation may be induced by selective advantage. Changes in climate may transform an area of succulent green plants to individuals better adapted to living under drier conditions, having thicker stems and better water-retention ability. This could be the result of immigration by more suitable plant species with the changing conditions. However, if the climate changed very slowly, selection might produce them. Feeding upon these plant stems there may be a population of voles. Some of these voles have a slightly better ability to conserve water, so that, despite the shift in climate, they are quite able to exist on drier material. The less adaptable will be pushed to the lower and wetter areas. Thus, a shift in the distribution of the two different strains will occur, leading, perhaps, even to physical isolation. Should they remain separated for a very long time, the genomes of these two strains will follow independent paths of random mutation and selection. Thus two distinct species may arise from one.

Divergence Over the Years

It is by continuous selection over millions of years that evolution has produced from a few lines of the primordial organism a kaleidoscope of different forms. It is a common misconception when dealing with evolution to envisage a missing-link organism that appears half-way between two organisms existing today. In other words, we look at today's frog and today's rattlesnake and look for an animal that is half frog and half snake. But the frog and the rattler are both organisms of today, and all organisms are changing with time. It may be true that at some point in time the frog and the rattlesnake had a common ancestor, but it is not true that by some miraculous process the frog as we know it today was transformed into the rattlesnake. Similarly, the monkey did not evolve into a human. At some point in time, an organism not unlike the monkey and not unlike the human, under the process of selective pressures gave rise to a line of monkeys and a line of hominids. If the selective pressures had been different, the products would very likely have been different. Today, the living organisms that we observe are like the tips of a tree, protruding into our time, but reaching back through billions of years.

About three and a half billion years ago, the proto-organism (which was to be the forerunner of all life today) emerged from the primordial soup. Reproducing itself again and again, this one-celled organism formed communities in various areas of the oceans. However, in the genetic material of this organism background changes were occurring, and these, combined with selective pressure from the environment, led to different varieties and strains of this organism, each tending to develop in its own way. At a point about three billion years ago, one strain became sufficiently different from the rest (shown as C in Figure 13-2). It parted ways completely from the rest and became more and more unique—for it would eventually give rise to the higher forms of life. Otherwise the slow selection of the proto-organisms continued, with more and more diverse varieties appearing, in the end giving rise to the bacteria of today. At a point about two billion years ago, line C again reached a branch point. Two major strains, D and E, arose through mutation and selection of this variety. D would eventually give rise to the modern fungal organisms, yeasts, moulds, and mushrooms. But E continued on its way, being selected and moulded by environmental pressures for another billion years. About one and a half billion years ago, an important split occurred in species E. Some organisms gained the ability to photosynthesize, and they became the green plants; others became the animal kingdom.

Each of these divergences marks the split of life into major categories: bacteria, fungi, animals, and green plants. Each of these divergences was followed by further ones. The bacteria split into the heterotrophs—those dependent upon other living organisms for their

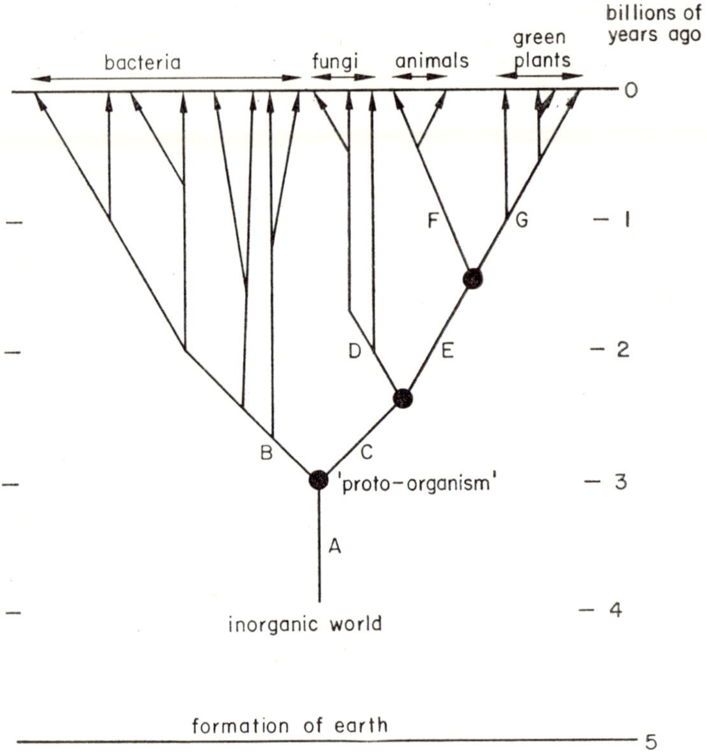

Figure 13—2 The Major Branches of Evolution

existence, and the chemotrophs—those dependent upon minerals in the environment, like the sulfate bacteria. The fungi also split up, into primitive forms such as yeasts and slime moulds which appear to have changed little over the many years of evolution, and into other forms such as mushrooms which are highly organized individuals. The plants, too, diversified, into algae, mosses, ferns, conifers, and flowering plants, and the animals into invertebrates such as the corals, the starfish, and the insects, and vertebrates such as the fish, the reptiles, the mammals, and the birds. This diversification into major groups of organisms is termed *phylogenetic diversification.* In this way the major phyla (or groups) of biological organisms parted from one another.

Phylogenetic diversification appears to have been initiated by important evolutionary events. These were mutational changes that over the years led to very strong selective advantages and that therefore offered new and promising avenues along which evolution could move. The organism, for example, that gave rise to the lines of modern animals and plants had gained something quite distinctive over the bacterial lines. It had acquired a mitochondrion, a unit that functioned as a highly efficient site of respiration. Its presence must have been very advantageous. Photosynthesis was another breakthrough of great advantage; so were the processes in primitive animals of laying down limestone-silicate shells (corals), of using blood to transport respiratory gases (ancestors of the starfish, molluscs, insects, and vertebrates), of the development of internal skeletal systems (in the chordates), lungs (in amphibian ancestors), warm-blooded physiologies (ancestors of the birds and mammals), and the production of womb-born young (ancestral mammals). (These are only a few of the many thousands of breakthroughs that occurred during evolution.)

Following each of these branch points, a characteristic burst of evolutionary activity occurred as these selective advantages were exploited in the appearance of many new species. Figure 13—3 illustrates some of these trends among the vertebrate groups. These

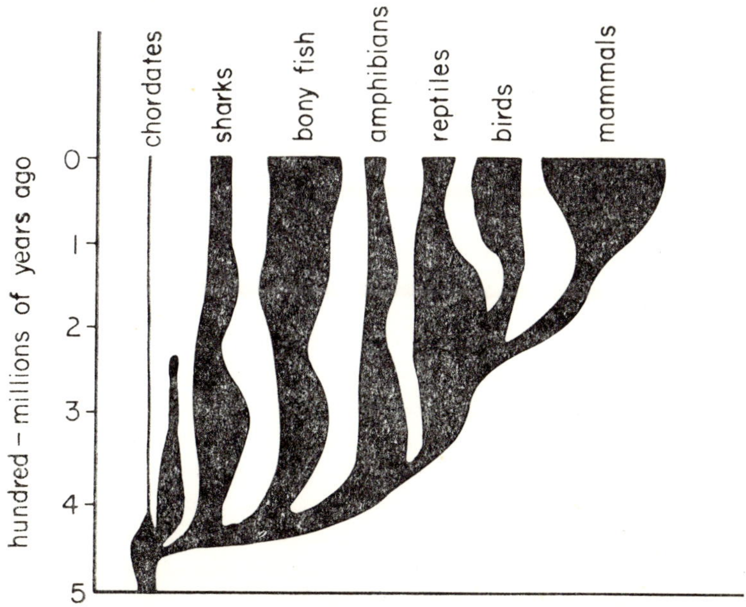

Figure 13—3 Phylogenetic Evolution among Vertebrates

trends, as reconstructed from fossil evidence, present a number of important characteristics of evolution. First of all, following a blossoming out (over millions of years) in each group of organisms, a diminution in the number of species occurred. This seems to have been compensated for by an increased diversity in another group. About 375 million years ago, the first reptiles diverged from the other groups of vertebrate organisms that were present. The species numbers of amphibians and sharks began to decrease. About 170 million years ago, the reptiles had reached their maximum. Dinosaurs roamed the earth, swam in the seas, and some like the pterodactyl even glided through the air. However, 125 million years ago the number of species of reptiles was rapidly diminishing. At the same time though, the mammals and the birds had begun to evolve, and the number of species in these groups rose sharply. This rise and fall in species number is integral to the second important feature of evolution. The lifetime of an individual species is limited. It has been estimated from deposits of bones and fossils that as many as 99 per cent of all the species that have ever existed are now extinct.

Extinction, however, is not a fast process in human time values. As we can see in Figure 13—3, it took a period of fifty million years for the number of species of reptiles to diminish, and even if a reptile lived twenty years on the average, in terms of single individuals this is an extremely long time! Extinction probably occurs as a result of many factors interacting together. Some bio-historians blame the fall of the dinosaurs on climatic changes. Others think that increased competitive pressure from the newly arisen mammalian lines was a factor.

The well-documented evolutionary history of the horse family is a good case in point. In total almost eighty species and numerous lines of this family have been identified, but of all of these only one exists today. Figure 13—4 schematically illustrates some of the widely diverse strains that developed and went extinct. (Each tongue represents a different line of evolution.) Each of these strains represents a different series of trends in equine evolution; some were trends towards smaller animals, others were the converse. Other changes occurred in dental structure, and in musculature and skeletal anatomy, but these trends have been found to be exceedingly complex. One trend however does stand out with some distinctiveness, and this is the movement from browsing to grazing.

Browsing organisms feed on herbs, on shrubs, and the lower

foliage of trees and other large plants. Their feeding habits affect the vegetative cover from one to several feet above the ground level. Grazers, on the other hand, feed predominantly at ground level, particularly on grasses. The ancestor of the horse some fifty-five million years ago, the *Hyracotherium*, was about the size of a fox, had some features in common with the modern-day horse, and was a browser. About forty million years ago, some of the descendants of this species were moving towards a grazing existence, and selection of dental features and of anatomy more suited to this existence was occurring. And about twenty-five million years ago, the two life-styles had begun to diverge widely with the formation of distinctive species suited to grazing and to browsing. Some increase in size occurred in the direc-

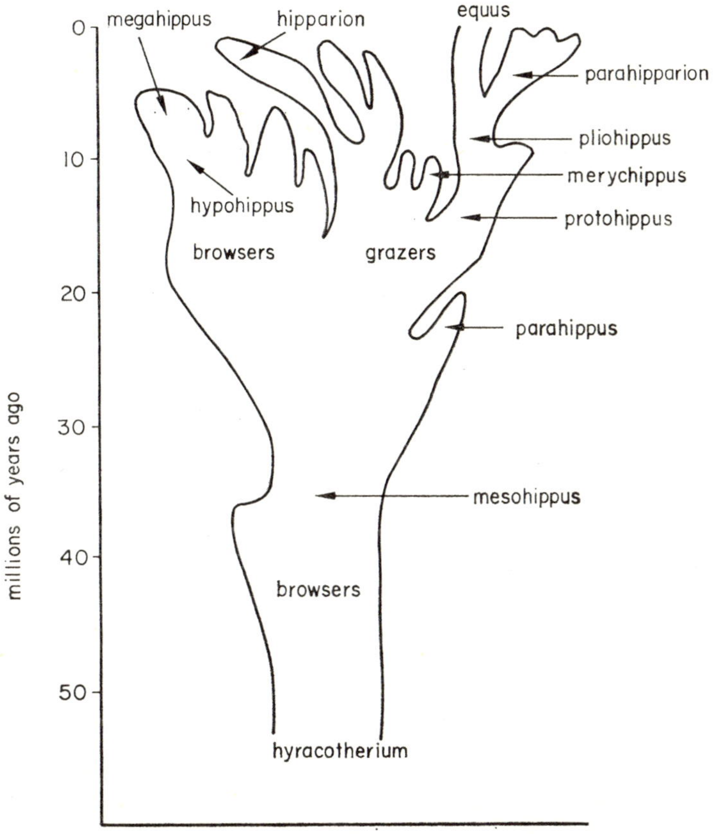

Figure 13—4 Evolution of the Horse Family

tion of *Mesohippus*, *Protohippus*, *Hypohippus*, but these increases were slow. Competition from both the grazing experts and the browsing experts pushed intermediate species to extinction. However, all the way through the Miocene period from about twenty to three million years ago, climatic changes were very slowly taking place. Rainfall in the temperate and tropical areas was decreasing, forests were shrinking, and grassland was increasing. Although this may have provided a stimulus to the browsers initially, eventually it was a factor leading to their extinction, for by two and a half million years ago, virtually all of the many browsing species had vanished. Only one line survived, that of *equus*, to give us the modern horse family.

Evolution and Life

The processes of life may be viewed on at least four different levels. First, the fundamental mechanisms of life operate within living cells. In fungi, plants, and animals, living cells aggregate in a concerted and self-regulating manner. This second level of life is the multicellular species. On a third level, living organisms act together in communities that are self-regulating and that influence the environment; these are termed *ecosystems*. On the fourth level, all of the ecosystems of this planet interact with one another and with the physical environment to form the biosphere—the spaceship of life. On each of these levels of life evolution has had different effects.

At the level of the cell, as we have already seen, evolution selected new and modified cellular structures and molecular processes. But over the four billion years of evolution, through the millions of organisms that evolved and became extinct, the most fundamental aspects of cellular mechanics were conserved. Today, for example, all cells share the same genetic code and very similar systems of protein synthesis. On the level of the cell then, evolution appears to have acted by playing variations upon a common theme.

Evolution has acted most dramatically at the level of the organism. While cellular processses were conserved, the conglomerations of cells that were living organisms moved through fantastic and wonderful shapes while adapting to a wide variety of environmental stimuli. Minute iceworms, for example, adapted to the cold of glaciers where they feed on organic matter deposited by the winds. In hot springs, certain forms of algae and bacteria thrive in near boiling temperatures. Organisms like the cactus have adapted to conditions of low water and sparse nutrients, while others like the salmon have developed physiologies for life in both salt and fresh water.

In the following two chapters we shall follow life on the level of the ecosystem and on the level of the biosphere. Here evolution appears to have had profound effects, acting upon the fundamental processes that characterized the very first living organisms in a manner that not only expounded upon them but also reinforced them. Out of a relatively simple system of a few species of living organisms participating as we have seen in the first processes of matter transformation in the environment, evolution appears to have created a highly complex and self-regulating web of life spanning the soils, oceans, and atmosphere of Earth. On a global basis this flow of matter has implications for all of life. In the next chapter we shall consider these implications, and in the following chapter we shall consider the interactions between living organisms today that stabilize the biosphere.

CHAPTER 1—4

Life and the Biosphere

Despite the great diversity of form that exists between them, all living organisms share a common denominator that reflects not only their common ancestry four billion years ago, but also the basic patterns of survival. Thus we can see that all living organisms function along three basic pathways forming a cycle, very much like the one of form-break-up-form that characterized the beginnings of life:

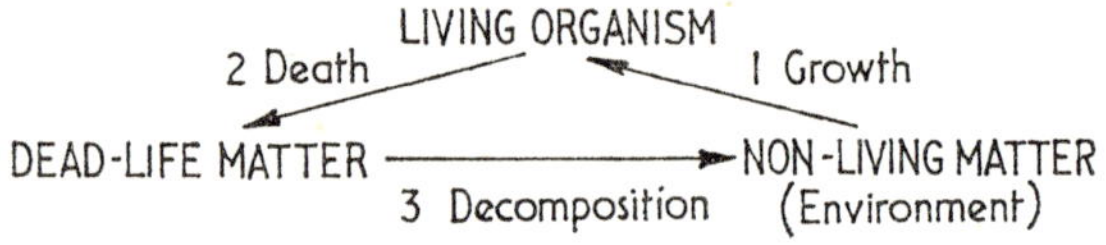

Figure 14—1

1. The first step of this cycle is growth, which is accomplished by taking in food matter from the environment. A grasshopper, for example, feeds on a wheat plant; a wheat plant feeds by absorbing minerals from the soils and carbon dioxide from the atmosphere. In both cases the primary source of matter lies in the environment as raw material such as phosphate, sulfate, or carbon dioxide.

Involved in this growth phase are all the energy-requiring processes of the living organism, including motion, perception, reproduction, and body maintenance. And for all these processes matter is assimilated and converted to body materials and the bonds in it are converted to useful energy and metabolic waste products which are released to the environment.

2. All organisms die at some point in their life history, most after reproducing their kind. Death may be brought about by any of a number of forces ranging from old age to consumption by a predatory animal. In every case the living matter contained in the body of the now dead organism becomes what could be termed dead-life matter. In dead-life matter the atoms, instead of existing in simple molecules such as carbon dioxide and nitrate as they do in the environment, are tied up in complex molecules of life forms. Because of death, however, these molecules fail to function as living beings.

The second step of the life cycle, then, involves all conversions of living material to non-living material, and at the end of this step there is a wide range of materials including dead organisms (plant, animal, or bacteria), feces, urine, carrion, and moulted skin, hair, skeletons, or shells.

3. Each of these materials is the starting point for the third stage of life, that of decay and decomposition. The process is carried out by a specialized and diverse group of organisms, the *decomposers*. Decomposers range from scavenging beetles and worms to yeasts and moulds and other fungi, and from protozoans to thousands of species of soil bacteria. The action of the decomposer organisms, in total, is a breakdown of the complex polymeric molecules of life to simple end products—carbon dioxide, nitrogen, and other nutrients.

As a result, all the matter initially assimilated by the living organism is eventually returned to the environment in its original form. Even the decomposers themselves follow these paths, growing, dying, and being decomposed by others of their kind. As a result we have the most important law of ecology: *There is no such thing as waste matter.* Waste is a species-oriented term, and what one organism does not want, another in the environment does. The implications of this are important.

First of all, many of the metabolic end products found in dead matter and in metabolic wastes, particularly nitrogen and sulfur containing compounds, are highly toxic to the organisms that produced them, and to many others as well. If these materials were not decomposed, they would accumulate in the environment, eventually reaching high enough concentrations to cause harm. Thus, the organism that secreted the waste in the first place would suffocate in its own poison as people have, in polluted cities.

Secondly, if none of this matter were decomposed, life would follow a one-way path. Assimilation and growth would steadily convert the nutrient raw material of the environment to dead-life forms or wastes. These would be unusable. Eventually, when the scarcest raw material ran out, for example phosphate, further life would be impossible. Thus, through this cycle, raw material is being constantly made available again to life processes.

Thirdly, by the long-term conversion of materials through the processes of life, an environmental transformation has occurred and is continually being maintained. In other words, the flow of elements through the processes of life in living organisms is maintaining the quality of the environment at and about the surface of planet Earth. This flow, for example, is regulating and maintaining the composition of the atmosphere. If we were to examine the flow of individual elements such as carbon, nitrogen, or sulfur through this cyclical pattern of growth, death, and decomposition, we would be following what are termed the *biogeochemical cycles*.

The Biogeochemical Cycles

These cycles are only in part biological, for they also contain powerful geochemical forces which are capable of transferring matter from one part of the planet to another. Forces such as wind and water erosion, volcanic activity, siltation and deposition processes, and the uplifting or sinking of land masses all have their effects. Some of these forces, erosion for example, are highly predictable ones, dependent upon rainfall and climatic conditions which are apt to change only over prolonged periods. Others, like volcanic activity, may be sporadic and sudden. Integrated with the geochemical forces upon the surface of the earth are the biochemical forces. The biochemical forces, as we have seen in the last chapter, have their roots in the very beginnings of life itself, and today, millions of years later, living organisms still carry out roles similar to those of their ancestors.

The Carbon Cycle

On Earth the element carbon is found in several major forms. In the oxidized form (atoms that are combined with oxygen are generally termed *oxidized*) carbon is found as carbonate rock (for example limestone or $CaCO_3$) and as carbon dioxide. Because it is very soluble in water nearly 98 per cent of all the carbon dioxide on Earth is found in the oceans. A small amount is also found in the atmosphere. In the reduced form (*reduced* is the chemical term for atoms that are combined with hydrogen), carbon is found either in living organisms, in the soils, or as fossil fuels.

The link between the oxidized and the reduced forms of carbon is life, with its complementary processes of respiration and photosynthesis. (In general the chemical conversion of the reduced form of an atom to the oxidized form is an energy-producing process. This then is the essential reason that life operates by oxidizing reduced carbon.) Figures 14—2 and 14—3 depict this process. Through photosyn-

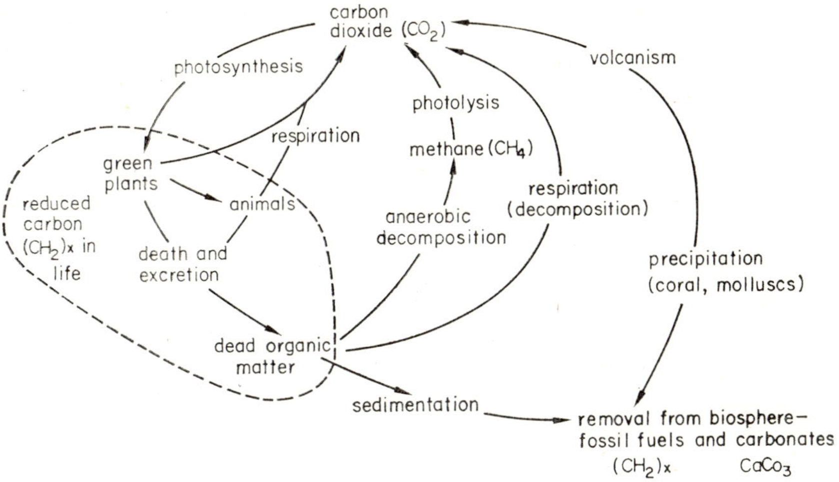

Figure 14—2 The Carbon Cycle

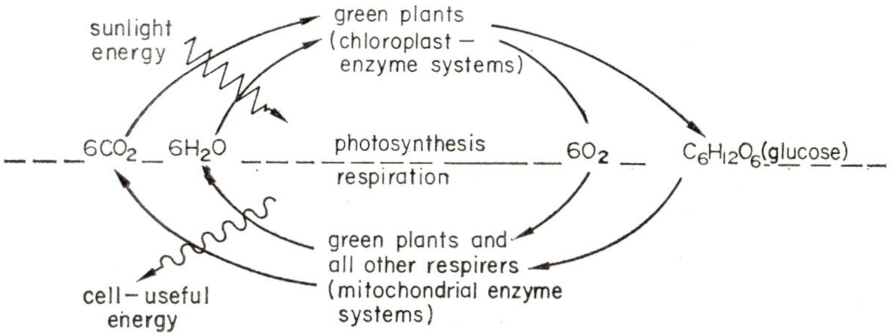

Figure 14—3 The Photosynthetic Balance

thesis sunlight energy is employed to reduce the carbon atoms in carbon dioxide molecules by linking them together and combining them with hydrogen atoms taken from water molecules. Glucose and oxygen are produced. The sunlight energy is thus stored in the bonds between the reduced carbon atoms of the glucose molecule. Through the reverse process of respiration, oxygen is consumed and water and carbon dioxide are produced. As the glucose molecule is oxidized the energy in its bonds is made available to the metabolic systems of the cell. Thus the carbon dioxide cycle is completed and the carbon dioxide that was initially fixed during photosynthesis is returned to the atmosphere. Respiration is essentially the same in green plants, animals, and many bacteria. However, in the absence of free oxygen some bacteria (termed *anaerobes*) can respire, producing methane or CH_4 instead of CO_2. This natural (or marsh) gas seeps slowly into the atmosphere and there, in the presence of sunlight, combines with oxygen to produce carbon dioxide, so that here too the cycle is completed. This anaerobic bacterial action may be a hang-over from the time when the atmosphere contained little free oxygen.

All of the carbon making up the bodies of animals and plants and composing the organic portion of the soils represents oxygen that has been released and has not recombined with carbon. Similarly, some dead plant and animal matter is continually leaving the system in the form of peat bogs, and lake and ocean sediments. Trapped away from oxygen, some of this material is slowly buried deep below the surface of the earth, without being converted back to CO_2. Consequently, this material accounts for free atmospheric oxygen. By far the greater amount of fixed carbon is returned as CO_2 during usage for energy purposes.

In the short term then, carbon is flowing from life to the environment and back again. In the long term, it has been the slow removal of carbon material from the biosphere that has produced an excess of oxygen in the atmosphere. This process presumably began with photosynthesis. By the end of the oxygen revolution 750 million years ago, atmospheric levels of carbon dioxide diminished to those of the present day. Yet in doing this, the biosphere has had to fix almost forty-five times as much carbon dioxide in the form of sugars as is necessary, simply to produce the oxygen that is present now. The reason that such an excess of oxygen is necessary is a simple one. Oxygen being a highly reactive element (even in the moderate condition of the biosphere) continually binds chemically to materials in the environment, for example to iron to produce iron oxides, to nitrogen to produce nitrogen oxides, and to sulfur to produce sulfates. Thus, without a continual production of new oxygen by the green plants, the molecular oxygen of the atmosphere would slowly disappear over a period of millions of years.

The Nitrogen and Sulfur Cycles

Like carbon, both nitrogen and sulfur exist on Earth in both reduced and oxidized forms. Oxidized forms of nitrogen include the nitrates and nitrites that are found in the soils, while reduced forms of nitrogen include the amino-nitrogen of proteins found in living organisms and the soils and the gas ammonia that is found in trace quantities in the atmosphere. Nitrogen is also found in the rather inert and elemental form of nitrogen gas in the atmosphere. Reduced forms of sulfur are also found in living organisms as well as the sulfides of rocks and ocean muds. Oxidized forms of sulfur such as sulfates and sulfites are found in the non-living environment, in rocks, soils, and dissolved in sea water.

The link between the reduced and the oxidized forms of these elements is again life. Processing between these forms is very much the result of microbial forces. Bacteria, fungi, and some algae of soils, bottom muds of lakes and oceans, and the subsurface regions of fresh and salt waters transform these elements from reduced to oxidized forms and back again in energy-yielding processes analagous to the photosynthesis and respiration of the higher plants and animals. These organisms occupy the regions of decomposition, and it is here where the rest of life has finished with biologically bound sulfur and nitrogen that the activity of these transformers is most intense.

With the aid of these powerful forces, sulfur and nitrogen flow through vast cycles in the environment. Nitrogen, for example, is fixed by bacteria such as *Azotobacter*, yeasts such as *Rhodotorula*, and, in damp soils and oceans, by blue-green algae such as *Calothrix*.

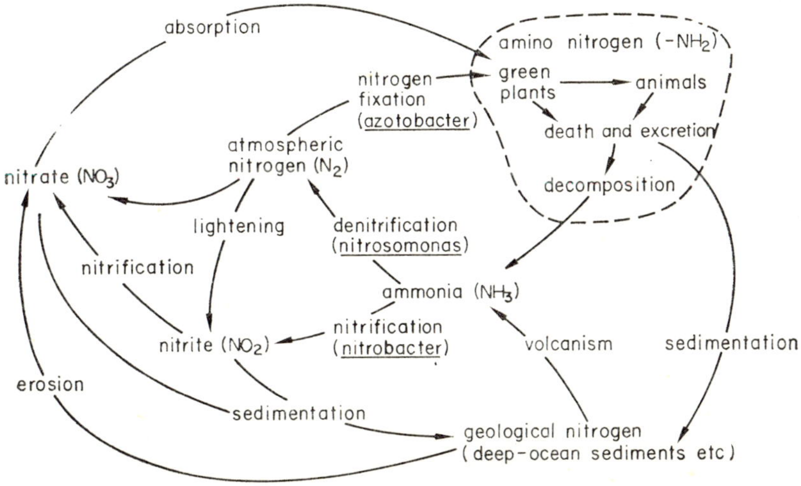

Figure 14—4 The Nitrogen Cycle

By converting atmospheric nitrogen to ammonia and to protein nitrogen these organisms make it available to the green plants. Certain of these nitrogen-fixers like *Rhizobium* attach themselves to the roots of leguminous plants, such as clover and alfalfa, in the form of tiny nodules. Here, in return for energy from their hosts, they carry out this essential process.

After it has been used by the rest of life, this protein-nitrogen can be released as ammonia by decomposers. Two possible fates await this ammonia. Either it can be reoxidized by bacteria such as *Nitrobacter* through the process of nitrification yielding nitrites or nitrate, or it can be denitrified by bacteria such as *Nitrosomonas* to atmospheric nitrogen. If it is converted to nitrates, these may be reabsorbed as nitrogen sources by either micro-organisms or by the green plants.

The flow of sulfur in the environment is the result of a similar battery of microbial forces acting along many complex and interlocking pathways. Both of these cycles have many geological inputs and outflows that supplement the biochemical pathways. Nitrogen and sulfur, for example, are removed from the biosphere as deep ocean sediments, and they are returned through volcanic activity as

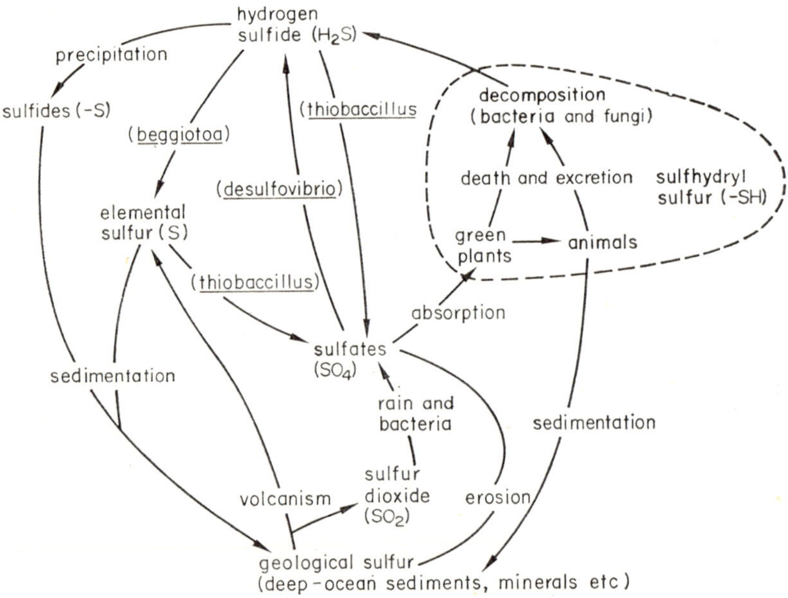

Figure 14—5 The Sulfur Cycle

ammonia, nitrogen oxides, and sulfur dioxide. High in the atmosphere lightning may convert nitrogen to nitrogen oxides, which come down in the rainfall as nitrates and nitrites. Together these biological and geological forces interact to balance the flow of sulfur and nitrogen in the biosphere, just as the flow of carbon and oxygen is balanced.

In the short term these flows of nitrogen and sulfur renew the fertility of the soils and provide living organisms with a continual supply of nutrient material. In the process, toxic nitrogen and sulfur waste products tend to be eliminated. In the long term all of the biogeochemical cycles are of tremendous importance. Let us consider them in this light.

The Long-Term Advantages of Nutrient Recycling

We have seen already how the oxygen revolution was attained; the advantages it brought were many. Among them it provided an atmosphere that facilitated respiration and the obtaining of energy, and a shield against high-energy ultra-violet radiation by the formation of an ozone layer in the upper atmosphere. Life was thus able to emerge from the waters to the land and the whole process of evolution that followed was made possible.

However, the long-term regulation of carbon dioxide levels has probably also been important, because the level of carbon dioxide in the atmosphere appears to be linked to the heat balance of Earth. Increasing carbon dioxide in the atmosphere would mean that more infra-red radiation would be detained, causing the atmospheric temperature to rise. More water would then evaporate and the water vapour content of the atmosphere would increase. And this could have two effects. First of all, water vapour is an even more efficient absorber of infra-red radiation than carbon dioxide, so the temperature of the atmosphere would rise further still. In addition however, increasing the water vapour content of the atmosphere will decrease its transparency and this could cut down the incoming sunlight, leading to a decrease of temperature and a reversal of the warming trend set into effect by the carbon dioxide. On the other hand, the atmosphere of Venus has a very low transparency to incoming radiation, reflecting 75 per cent of it away. Yet the effectiveness with which the atmosphere of Venus is able to trap heat radiation is postulated as the reason for its very high atmospheric temperatures. Thus, increasing water vapour levels in the atmosphere of Earth might very well lead to a more efficient trapping of heat radiation, even though less energy was entering in the first place, and subsequently the temperature could rise instead of fall. In the end, a new equilibrium of atmospheric temperature would be reached with its initial determining factor the amount of carbon dioxide in the atmosphere. Increasing carbon dioxide levels would push the system still further towards a new point of equilibrium. If the carbon dioxide content of the atmosphere were decreased, a similar process to that described might occur. A large number of other factors would also come into play. The oceans for example are a great heat sink, and a small rise in carbon dioxide could be compensated for simply by a greater absorption of carbon dioxide or heat into the vast oceans.

It might be pointed out that, from the burning of large quantities of fossil fuels, the human race has already changed the concentration of carbon dioxide in the atmosphere, and many scientists feel that we may already be affecting the temperature balance of the planet. (This will be discussed in greater detail in Chapter 2—5.)

There appear to be mechanisms in the biosphere to avert increases in atmospheric carbon dioxide. First of all, carbon dioxide is very soluble in water—almost 80 per cent of the Earth's carbon dioxide is dissolved in the oceans—and a slight increase could thus be simply absorbed by the oceans. Secondly, over long time periods the processes of life may be important. An increase in carbon dioxide and in world temperatures may, if carbon dioxide is the limiting factor, result in greater photosynthesis with the consumption of more carbon dioxide. Warmer temperatures would mean also a slight warming of the oceans, and an increased activity in coral and shellfish organisms. These creatures lay down great amounts of carbon dioxide in their limestone (calcium carbonate) skeletons, and many of the world's limestone rocks are a result of such activity. In the past these organisms may thus have played very important roles as the atmospheric carbon dioxide levels decreased to the present ones.

All of the biogeochemical cycles may be linked together by processes occurring in sea water. Figure 14—6 shows a typical cross-section of sea water, with the two 'end points' of biogeochemical cycling—the producers and the decomposers. The producers are generally located in the upper regions of the water where light penetration is adequate to support photosynthesis. The organisms feeding

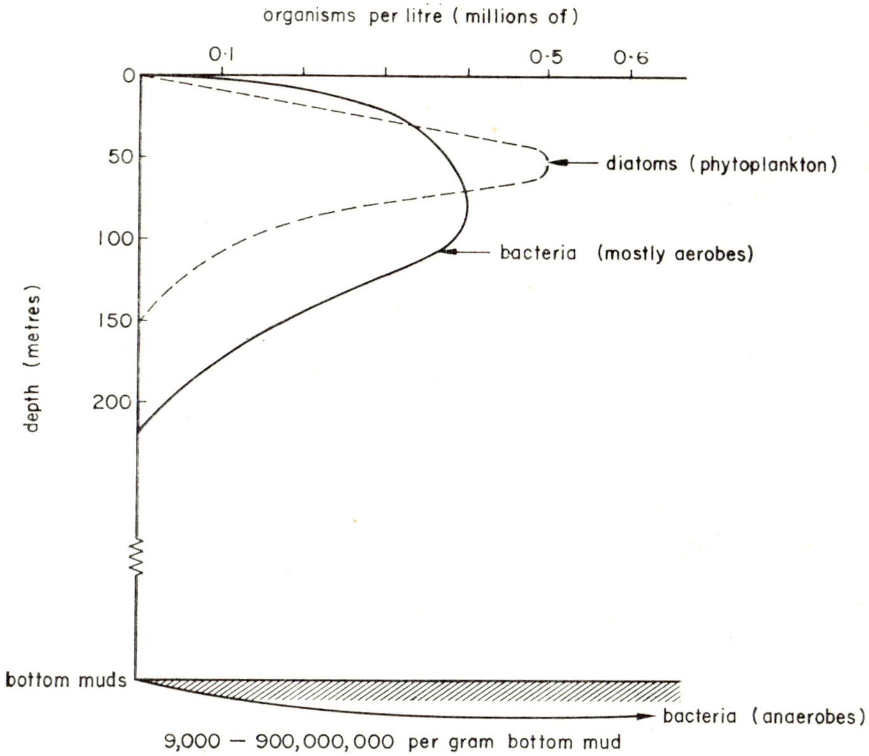

Figure 14—6 Oceanic Ecosystems

on the producers inhabit this level and the ones below it. From these levels the wastes of the producers and their consumers sift downwards. Thus the decomposers tend to be concentrated at a level below the producers as well as on the sea floor—the ultimate resting point for detritus from the upper levels. In the upper reaches, the decomposition processes tend to be aerobic ones, because dissolved oxygen levels are fairly high and replenished by turbulence; at the bottom they tend to be anaerobic. Both of these mechanisms seem to have a specific importance.

Since phytoplankton are the producers of the oceanic food chains, the ratios in which they take up nutrients such as carbon dioxide and nitrate from the sea water tends to determine the ratios of these nutrients in the final stages of decomposition. That is to say, when all of the organic matter is completely decomposed, the nutrients in it will be released in the same proportions as in the original fix. Phosphorus appears to be, over all, the rarest nutrient in sea water, and the amount of phosphate will therefore limit the amount of the other nutrients taken up.

Table 14—1 illustrates this point in another way:

Table 14—1 Nutrients in Sea Water and in Phytoplankton

	P	N	C	O
1. Elementary analysis of phytoplankton	1	16	106	−276
2. Elements available in sea water	1	15	1000	200–300
3. Changes observed in sea water	1	15	105	−235

For every phosphorus atom in phytoplankton there are 16 nitrogens and 106 carbons. But there are available in sea water, for every phosphorus, 15 nitrogens and 1,000 carbons. The suggestion has been made that over many years life in the oceans has regulated the phosphorus and nitrogen ratios. However, more striking is the relationship of these ratios to the actual concentration of oxygen in sea water. If one were to carry out the complete decomposition of the phytoplankton, thereby converting all of the carbon to CO_2, nitrogen to NO_3^-, sulfur to SO_4^{2-}, then one would need 276 atoms of oxygen. And dissolved in one litre of sea water there appear to be between 200 and 300 oxygen atoms for every phosphorus. Line 3 of Table 14—1 gives the observed changes in sea water, and the correlation is striking. Less oxygen is consumed than predicted because some of the decomposition occurs anaerobically in the bottom muds.

It has been suggested, therefore, that an equilibrium determined by the amount of phosphate in sea water and the solubility of oxygen has been set up by life in the seas. The solubility of oxygen is dependent among other factors upon its concentration in the atmosphere, around 20 per cent. The equilibrium has reached the stage where the bio-mass produced by the phytoplankton can be recycled by oceanic ecosystems because there is just enough oxygen available to do this.

In the depths of the oceans more subtle processes seem to be at work; Figure 14—7 illustrates some of these. Here, where oxygen levels are very low, nitrate and sulfate-reducing bacteria are at work. These organisms use the oxygen that is combined chemically in nitrates and sulfates in order to respire; they do it by combining this oxygen with organic matter and releasing carbon dioxide. As carbon

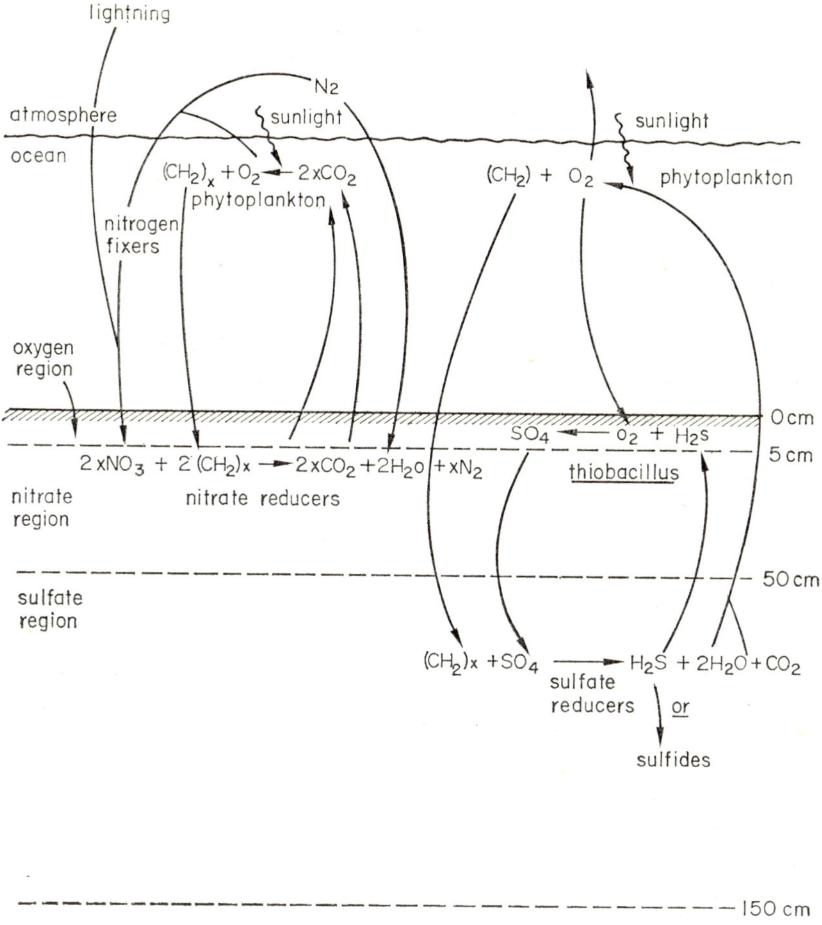

Figure 14—7 The Ocean Floor Cycles

dioxide percolates upwards it becomes available for photosynthesis again. In the end, a net production of oxygen occurs, because carbon dioxide is produced without the consumption of free oxygen. However, many of these nitrates and sulfates were originally produced by oxidation of nitrogen or of sulfides on the land masses. These continental processes are thus continually being reversed by organisms such as these in the ocean muds. In this way, these organisms are reinforcing the photosynthetic action of the green plants.

All of these processes have taken millions of years to bring about.

Without the sulfate and nitrate-reducing bacteria on the ocean floors, for example, the atmosphere would revert back to an oxygen-deficient state in about ten million years. That is certainly a long time, but it is indicative of the continuous and long-term scale on which the biosphere operates. Considering that the majority of life-forms as we know them evolved in the last 700 million years, the time for the atmosphere to revert back is not really so very large.

Energy and the Biosphere

Matter, we have just seen, circulates about the biosphere in great cycles. Energy, on the other hand, moves in a different manner, by cascading through the biosphere to an irrecoverable form called waste heat.

The flow of energy obeys two fundamental laws of physics from which there is no escape. First of all, according to the Law of Conservation of Energy, energy can never be created or destroyed. Energy coming into the biosphere is, therefore, reflected away unchanged or else transformed into other forms of energy. For example, sunlight, on hitting a plant leaf, may be reflected (the green wavelengths), or it may be absorbed (the red-yellow bands) and transformed into the chemical energy of carbon–carbon bonds. As energy is released during the metabolism of these bonds it follows similar rules. It may be transformed into kinetic energy of motion, into light energy (for example, the fire-fly), electrical energy (nerve impulses, the electric eel), or into many other forms of chemical energy, but it is always conserved.

According to the Second Law of Thermodynamics: no process can ever be 100 per cent efficient. In other words, at every stage of the transmission, transformation, or usage of energy, a certain percentage of the total is lost or dissipated in an irrecoverable form. The form in which it is usually lost is termed *waste-heat energy*. This form of energy consists of minute amounts of infra-red radiation that are diffused over a large area. Let us examine, for example, the contraction of an arm muscle. Within the chemical processes that actually contract the muscle fibre, at the levels of tendons, at the joints of the bones, and in the actual work carried out by the muscles, there are basic inefficiencies which can be minimized (and are minimized by living systems) but never overcome. At each stage waste heat is produced. It is expressed for example as the friction that occurs during movement of the bone socket. This wasted energy is dissipated in waves throughout the surrounding medium. Moving outwards, this waste heat sets up vibrations in molecules in its path, these in turn knock other molecules, setting up lesser vibrations, and so on. As each of these molecules vibrates, it loses minute amounts of heat energy before it settles back into the normal state. Thus, eventually the flesh about the muscle warms up and the clothing about the flesh. Following each of these steps the energy becomes more and more diffuse, until finally it reaches such low intensity that it has little further effect.

This is the fate of all energy—in space, in the solar system, and in the biosphere. At every stage of energy-use and transformation the proportion that is useful decreases, until in the end all of the energy has been dissipated as waste heat. For this reason, energy has been described as cascading through the biosphere, being used at all stages of life from the green plants to the decomposers, but at every level of this use becoming partially or totally converted to waste heat. It is, therefore, only because Sun provides the biosphere with a constant stream of radiant energy that life can turn its cycles. Thus life employs matter and energy in two very distinctive ways, and these differences could be termed the Second Law of Ecology.

Energy and Living Organisms

Energy is employed at all the levels of life from within a single organism to the biosphere as a whole. It is used in the breaking and forming of chemical bonds during the process of building body units; it is used in the process of motion and of nervous sensations; it is used for body maintenance and regulation and in the process of digestion; it is used during cell division and reproduction.

As we have seen, the energy used by life is trapped from the radiant energy of Sun by the green plants during photosynthesis. Strictly speaking, there are some micro-organisms that obtain their energy from processes like the oxidation of sulfur to sulfate and the oxidation of nitrogen to nitrate. Other bacterial organisms obtain their energy, as we have seen, by using the oxygen combined with sulfur in sulfate molecules to transform hydrocarbons like methane to carbon dioxide. The first are dependent upon an oxygen-rich atmosphere and thus upon photosynthesis. The second type require adequate amounts of reduced carbon, and again this is largely dependent upon photosynthesis. Had the oxygen revolution not occurred, and should these primitive energy-gleaning mechanisms have been the only ones persisting, then simply from shortage of material alone these processes would probably have exhausted themselves millions, if not billions, of years ago. We can safely say then that photosynthesis is the prime mechanism by which life receives its needed energy.

In plants, animals, and bacteria, the carbon-carbon bonds of the glucose molecule are split again as energy is needed, and the carbon dioxide units released. Many of the glucose units, however, are not immediately used for energy purposes. Some of them are chemically modified and linked together with other modified glucose molecules in a vast number of diverse ways. Many of these combinations can only be effected by specific organisms, while others can be assembled by nearly all living things. Green plants, for example, can link the glucose molecules end-to-end by a specific linkage. The product, vastly different from the sweet-tasting and water-soluble glucose molecule, is a tough and water-insoluble molecule called cellulose, which clumped together with others of its kind forms the structural backbone of cell walls of plant leaves, stems, and roots, and thus of wood. On the other hand, plants and some animals can link glucose molecules together in another way, to produce starch. Or they can transform glucose into a molecule called *α-Ketoglutarate* and combine this with ammonia to give amino acids—all in a series of enzyme systems. Lipids can also be produced, or carbohydrates such as fructose or sucrose (cane sugar), or complex molecules like hemoglobin or chlorophyll. (Figure 14—8 illustrates these processes.)

However, at any point the process may be reversed. Amino acids, lipids, hemoglobin—all of these can be broken down again and their carbon-carbon bonds split to produce carbon dioxide and the energy contained in them. This does not occur universally, however. Insects and spiders, for example, are covered by a tough exoskeleton of chitin, a material that is a polymeric assemblage of N-acetyl-D-glucosamine (a sugar) subunits. Birds are unable to digest the chitinous portion of their prey and it is excreted with the other indigestible materials. Some bacteria, protozoans, and a few invertebrates like the earthworm, however, have the ability to break chitin down to its glucosamine subunits which can then be metabolized for energy and building purposes.

If we were to follow the movement of matter and energy from the green plant to the animal feeding on it, the antelope for instance, and from the antelope to its predator the lion, we would be defining what

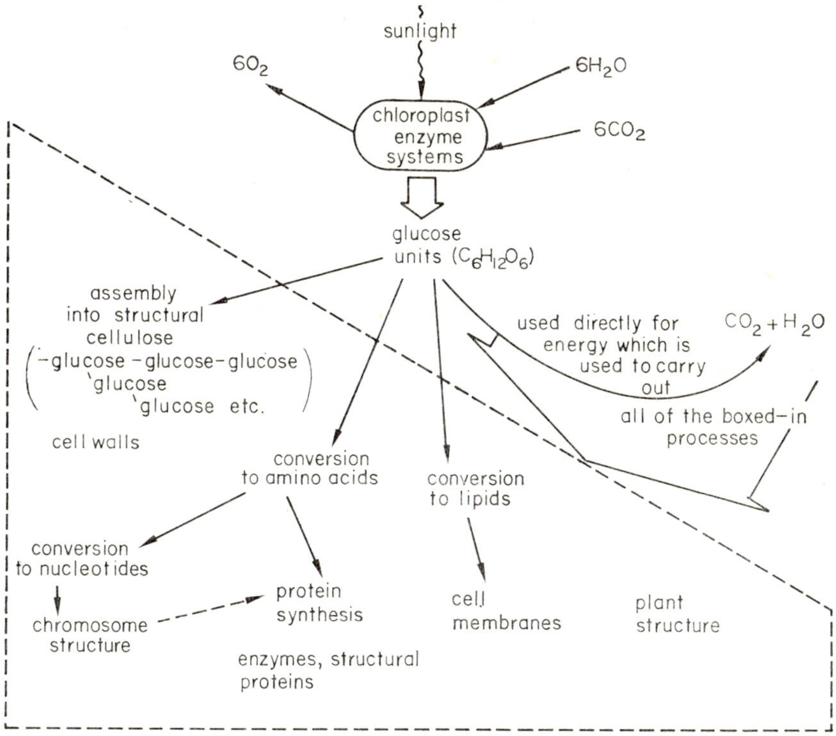

Figure 14—8 The Uses of Glucose

is termed a *food chain*. A general picture of food chains is important in summarizing the cycling of nutrients. It is through these chains, from producer to herbivore to carnivore to decomposer, that matter flow is maintained and regulated, and environmental quality maintained. Each food chain acts as a link in the biogeochemical cycles; and the total action of food chains all over the planet is the total action of the biogeochemical cycles.

The use of glucose at every level of the food chains is the driving force of the biosphere. The energy released is the force behind the consumption of matter by living systems, its assembly into subunits and the polymers of life, and their eventual decomposition again at the end of the cycle. The formation and metabolism of carbon-carbon bonds by living organisms is at the basis of all life and upon it life, the biosphere, and their stability ultimately depend.

Every organism, from phytoplankton to shrimp to musk-ox, is really a node in these cycles. The amount of flow through that node is dependent upon the size of the node, and there exist intricate mechanisms, as we shall see, to regulate the population of each organism and thus the size of each node. The interactions between living organisms thus effect the biogeochemical flow of matter through the biosphere and are therefore essential in maintaining long-term environmental quality.

CHAPTER 1—5

From One Organism to the Next

The human brain, so frail, so perishable, so full of inexhaustible dreams and hungers, burns by the power of the leaf.
Loren Eiseley, The Unexpected Universe, 1969

We have seen that both on the global level of the biosphere and the individual level of the cell, energy moves the biosphere. Because of it, matter flows through living organisms and through the biogeochemical cycles with their all-important environmental consequences. Through energy, every individual of every population is able to carry out the processes and functions of its life. Among these living organisms there exists a series of complex interrelationships for the transfer of this energy which have been described, in a simplified way, as food chains. The study of these interrelationships is a major part of the science of ecology. How do individuals of two different species interact together, if at all? How do they interact with hundreds of other species? What is the effect of the sum total of these interactions occurring together in a community of organisms? How do these interactions vary with the population sizes of prey and predator organisms, with inherent biological instincts, and with competition within the population of one organism? In total, how does energy travel through communities of living organisms and how much of it is lost along the way? All of these questions are vital ones, for they determine the way in which matter flows through the living communities, each an essential part of the total biosphere.

An Organism and an Ecosystem

A frog is a good animal with which to begin this discussion. It is an amphibious animal, adapted to living part of its life on land and part in water. The frog's bodily metabolism will allow it to live only under certain conditions of wetness, salinity, and temperature. Not all of these limits are easily definable; for example, the frog may be able to survive over a whole range of temperatures, the extremes of which vary with other conditions. We may find, for example, that the frog can survive warm moist conditions but not warm dry ones. Or it may survive the latter only when it is well fed. However, we know that there are limits to the conditions under which a frog can live. These limits best describe the habitat of the frog.

During its adult life, the frog feeds mainly on insects such as flies and beetles; it is thus a *carnivore* or a meat-eating organism. But during its early life as a tadpole, the frog is a *herbivore* feeding off plant matter. The adult frog, being limited by such innate factors as eyesight and hunting methods, is capable of catching and ingesting only certain types of food. If we were to look at the composition of the frog's diet, we would find it composed more of certain insect types than of others. It might include, for example, an occasional dragonfly or earthworm, but for the most part, it would consist of mosquitoes and flies.

The frog's feeding habits, then, have a specific impact upon the other organisms of the surrounding community. Furthermore, the frog itself acts as a good source of food for many other organisms such as the pike, the heron, and the raccoon. The ability of the frog to escape predation is also linked to its basic physiological nature through speed of perception and motion, and through camouflage. In the end these determine the survival ability of the frog as a species and the impact that the frog has on the surrounding community as both predator and prey. This impact upon the rest of the community and upon the local environment defines what is ecologically termed the *role* of an organism.

Together the intrinsic limits of an organism define both the habitat in which that organism can live and the role that it can play within that habitat. Ecologically they also help to define a concept termed *niche*. If a habitat could be defined as the address of an organism, then the niche is the profession of that organism. In other words, the address of the frog is an area that is temperate and provides adequate dampness or wetness. The profession of a frog is a small vertebrate herbivore (during youth) and carnivore (during adulthood) that inhabits damp locations and provides food for certain mammals, birds, reptiles and fish.

The ecological niche of each organism places it in a specific location within a series of food chains. We have already considered the concept of a food chain—a series of organisms, one feeding upon the next, as the fly feeds upon dead plant matter, the frog upon the fly, and the heron upon the frog. Such an interpretation, however, can grossly oversimplify the actual picture. For the frog feeds not just upon the fly but also upon the mosquito, the earthworm, the water strider, and the dragonfly, among others. And the heron feeds upon small fish, frogs, salamanders, toads, and even crayfish. Thus the picture is more complex than a single food chain would lead us to believe. For this reason, the *food web* is a better concept to employ. Consider then, among a community of organisms, all of the possible food chains running from one organism to another. In total, all of these chains constitute a food web.

We will turn to the green plants as a focal point of these webs. Because the green plants fix the sunlight energy needed by all living organisms, they are the point of inflow of energy into the system. At the same time they fix all the nutrient matter needed by the rest of the system. The green plants are thus defined as *producers*, and it is no surprise that all of the rest of life tends to be arranged in webs about the green plant communities.

What is it that determines the nature of the producer communities? Here it is very much the limits imposed by the physical environment, although these in time may be altered by the plant communities themselves. These limits may include the amounts of available nutrients like copper and molybdenum, the amount of sunlight for specific periods of time (the growing season), and the amount of rainfall—all limits that partially define the niches of various plant species. And therefore, dependent on the conditions, we will find plant communities that are highly adapted to survival under those conditions—whether the cactus in a desert or the coniferous tree in the forest. All these factors will affect the subsequent environment—the amount of energy fixed and available to other organisms, the amount of shelter from wind and sunlight, the number of nesting areas, and so on. They will therefore play an important role in determining the types of organisms likely to be assembled about these plant communities.

The herbivorous organisms that feed off the green plants are termed the *primary consumers*; they feed first-hand off the food-base of the biosphere. Feeding on these consumers are the carnivores, the *secondary consumers*. Feeding on both the primary and secondary consumers are the *tertiary consumers*. Above this level there may even be *quaternary consumers*. Food webs rarely extend beyond this level. Consuming the wastes from each of these levels are the familiar decomposers. Each of these levels, from producer to consumer to decomposer, is termed a *trophic* level.

The arrangement of organisms around a plant community in this way, along with the associated areas of non-living but biologically maintained environment such as the soils or river and lake muds is termed an *ecosystem*. The absolute limits of any ecosystem are difficult to define, because some organisms play a part in more than one ecosystem. A forest ecosystem, for example, may be based on a widespreading cluster of hardwood trees. Surrounding this forest on one side may be grassland, on the other side a river community, both ecosystems themselves. A fox may spend most of its time in the forest hunting rabbits, but occasionally may wander out onto the grassland in search of a few ground-hogs. Similarly, a hawk in the forest may feed predominantly on the many insect-eating birds there, but occasionally may have a go at the small birds on the bank of the river. An insect living in the forest may be consumed by a warbler, or, alternatively, it may be washed by a rainstorm into the river where a trout gobbles it up. In the forest itself there may be patches of boggy areas with a different type of plant growth from that elsewhere in the woods. An ecosystem will be assembled about these small bogs. An ecosystem may even by assembled about a rotting log in the forest, within the decaying matter of the soils, or about a group of pines growing on a unique stretch of sandy soil within a predominantly maple forest. Each of these ecosystems may have unique organisms, while between them they will share others. It is therefore difficult in reality to delineate the boundaries of an ecosystem. Yet this does not detract from the value of the ecosystem concept in aiding in the understanding of living systems. And perhaps as we further explore the relationships within these communities of living organisms, we shall glean a better conception of the boundaries of an ecosystem.

The producers at the base of an ecosystem are widely varied from one area to the next. They may be phytoplankton floating in the upper layers of lakes and oceans, or they may be colonies of plants fixed to the bottoms of shallow lakes and shorelines, both areas where sunlight penetration is enough for photosynthesis. They may be the plants of a marsh, of a grassland, or of an equatorial rain-forest. They may be an alpine meadow or a coniferous forest. Primary consumers likewise are varied: shrimp feeding on phytoplankton, deer or moose feeding on grass, shrubs, and the lower branches of trees, seed-eating birds such as the pheasant, or leaf-eating insects such as the tent-caterpillar. Secondary consumers may be frogs, skunks, trout, herring, or starfish.

Tertiary consumers are usually those that appear to top the food chains; in other words there are no carnivores above them. These *top carnivores* include such organisms as the lion, osprey, pelican, pike, and wolf. Strictly speaking, there are often organisms feeding from this level, but these organisms tend to exist by parasitism rather than direct kill. Quaternary consumers therefore include such animals as the flea, tick, liver-fluke, and tape-worm.

We have mentioned only a few of the hundreds of thousands of species that fill out the food chains of ecosystems about the biosphere. And it is not always the easiest thing to determine the position of an organism in a food chain. A bear, for example, can be a primary consumer (feeding off berries), a secondary consumer (raiding birds' eggs), or a tertiary consumer (catching salmon), or very occasionally a quaternary feeding on humans. This problem has been resolved by placing the organism in its role furthest from the producer. In other words, the bear is a top carnivore, feeding on all the levels below it, but rarely preyed upon itself. Ecosystems do not succumb easily to analysis, and definition in this way of the various roles of each organism is useful in attempting to uncover the mechanisms by which the system operates, and the degree to which an organism affects the total system. It is obvious, however, that the more roles an organism takes on within an ecosystem, the more difficult it is to determine the total impact of that organism.

Between ecosystems there is a strong parallel of structure. Ecosystems, whether based on grasslands or marine phytoplankton, appear stacked upwards from the producers through trophic levels of primary, secondary, and tertiary consumers, all feeding into the decomposer level through the processes of wastage and death. The reason for these similarities is a fundamental one; it probably lies within the function of the ecosystem in transferring energy from one trophic level to the next.

Populations, Energy, and Food Webs

The organisms at each trophic level employ energy for their own needs. The green plants require a certain amount of the energy that they fix, for building new leaves and roots, for stem growth, for seed production. A rabbit feeding on the grass needs energy for growth, maintenance of body temperature, reproduction, motion, and so on. The same applies to the hawk and to the tick on the hawk's wing.

Thus, as one moves up a food web there is less and less energy available, simply because some of it has been used along the way. Two other factors reduce still further the amount of energy that passes up the food chain.

The structures of the green plants represent energy that has been trapped in photosynthesis. Should the rabbit consume all of the green plants, these plants would not survive. A certain amount of the energy fixed in their structure is necessary for their continued survival. Similarly, a certain number of rabbits have to survive if the rabbit colony is to maintain itself. The necessity of keeping a certain amount of the energy at each trophic level therefore reduces the amount that can be transferred to the next level. Finally, according to the Second Law of Thermodynamics, all of the processes in which energy is used and transferred are inefficient. . . . This cuts back substantially the amount of energy available for transfer at each level.

Because further along a food web less energy is available, trophic levels other than the decomposers rarely exist beyond the quaternary consumers. Even though living organisms have maximized their efficiency at capturing and utilizing energy, this restriction still holds. Furthermore, ecosystems tend to be broad and interlinked. We find that primary consumers, for example, spread themselves across many different food sources. Deer concentrate on grass and leaves, the porcupine on tree bark, and the rabbit on new shoots and small plants. Some organisms like the spruce budworm are almost solely restricted to one of the many possible food sources at the base of the ecosystem. Particularly at the level of the top carnivores feeding habits are very diverse, spreading across all trophic levels. Here energy is hardest to obtain and the consumers at this point spread themselves wide in an effort to get it.

Table 15—1 shows the energy budget of the aquatic ecosystem of Silver Springs in Florida. First, notice that of all the light absorbed by the green plants, less than one-twentieth is actually fixed into the form of biologically usable energy. Notice that more than half of the energy fixed is used by the green plants themselves through respiration for their own bodily maintenance and reproduction. Then notice how the energy transferred from one level to the next goes down by a factor of about ten each time, starting with a net production of 8,833kcal/m^2/year in the producer level, but only 21kcal/$metre^2$/year of this reaching the top carnivores. Here then are the physical limitations that exist on the number of trophic levels within an ecosystem.

Table 15—1 Energy Flow in an Ecosystem, Silver Springs, Florida*

Trophic level	Incoming energy†	Energy stored in bio-mass	Used for own purposes, or wasted	Energy passed to the next level
Solar radiation	1,700,000	—	1,290,000	410,000
Green plants	410,000	8,833	401,167	3,368
Primary consumers	3,368	1,478	1,890	383
Secondary consumers	383	67	314	21
Top carnivores	21	6	15	—

*Adapted from a study done by H. T. Odum, Ecological Monograph No. 27, pp. 55-112, 1957.

†All energy values are in kilocalories per $metre^2$ per year.

The figures given in this example are peculiar to the Silver Springs ecosystem. They will vary in many instances in other ecosystems. For example, it is obvious that the amount of sunlight will affect primary productivity. Transfers between some trophic levels may be more efficient in other systems, while other transfers may be less so. However, in most ecosystems transfer from one level to the next is about 10 per cent of the supplying level.

There is yet another reason why ecosystems are so similar in structure, and that reason lies within the roles of various organisms in maintaining its stability. Let us determine the means by which this occurs by looking at a typical ecosystem, a temperate forest ecosystem. In Canada such an ecosystem could be found in regions of good rainfall and moderately long growing season. For example these forests once covered much of southern Ontario, areas of the Eastern Townships and the St. Lawrence River valley in Québec, and the Fraser River valley in British Columbia. To a large degree this depiction will be oversimplified—there are many organisms and relationships that have been left out, indeed it is safe to say there are probably quite a number of these that are at present undetermined. Perhaps, however, by looking at the general picture we can draw together some of the concepts of a functioning ecosystem.

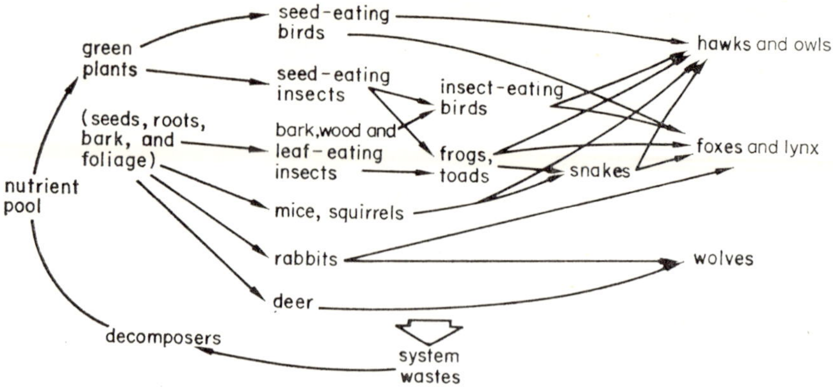

Figure 15—1 Food Web of a Temperate Forest Ecosystem

The simplified ecosystem is shown in Figure 15—1. The plant communities that form the base of it are varied and present a heterogenous appearance. Stands of maple, oak and elm hardwoods are interspersed with stands of conifers like white pine, or in boggy areas, cedar. Occasionally there are patches of new growth following recent clearing or fire; here one finds shrubs, and birch and alder. The variety of plant types and shapes provides a multitude of environments varying in sunlight, moisture, and wind cover, and providing opportunities for hunting or hiding, depending upon the organism.

Notice first of all the feeding habits of various organisms. At the lower levels, feeding is highly specific. Rabbits feed on shoots and small plants, deer feed on the lower leaves of trees, and grassy matter. At the higher levels, feeding becomes less specific. The fox, for example, can glean its food from mice, rabbits, snakes, grouse, birds' eggs, and in times of intense food shortage even berries or grasshoppers. As we have already seen this reflects the energy scarcity at the top of the ecosystem. But this non-specific feeding has another function so to speak. It provides stability to the food web. Should, for example, the myxomatosis virus wipe out a large part of the rabbit population, thus pushing it near extinction, the lynx will shift his feeding to a more predominate organism, say the young deer or the wood-grouse, thus giving the rabbit population a chance to recover and reinsure one of the lynx's food sources. (This however is not a conscious shift, simply a result of the lower availability of rabbits.) This is only possible because the lynx has non-specific feeding habits.

The size of the populations of organisms at every level of the food web is thus a very important thing. Should a population of leaf-eating insects grow out of control and decimate the leaves of the forest, the effects will be widespread. Not only will the environmental nature of the forest be changed (through more sunlight reaching the ground and loss of protective cover from the foliage), but the food supply of many other organisms dependent upon this vegetation will be affected. In time even the leaf-eating insect itself will run out of food, and face starvation as a result. Thus it is to the benefit of the leaf-eating insect itself that its population be controlled. This applies equally to any organism at any point in the ecosystem. Should a top predator such as the owl outgrow in number its food supply, then the owl's prey will suffer and so will the other top predators with whom he shares this prey. Furthermore there will be a chance of the lower trophic levels like the plant-eating insects growing out of control.

The control of population size in any group of organisms is thus a very important thing. Without it, anarchy would reverberate through the ecosystem as populations outgrew their food supplies to the point of endangering them. Six fundamental mechanisms seem to be at work keeping populations under control in an ecosystem.

1. CLIMATE Periods of heat, drought, and severe cold all take their toll, particularly on the old and the weak, but also on any organisms not adapted to surviving these conditions. Hurricanes, violent storms, floods, have a more general pruning effect.

These climatic factors tend to be fairly independent of the population density of the organism. For an adequate system of population control, one needs controlling factors that are responsive to increases and decreases in the size of the subject population. Some of the other mechanisms of control, like disease, predations, competition, and the availability of food act in this manner.

2. DISEASE AND PARASITISM Diseases are common to forms of life from all parts of the biological world. Parasites (organisms that live on others without killing them) are also common. Plants such as bracket fungi and mistletoe live on other plants; others such as rusts and scabs infect fruits and leaves. Many forms of yeasts and moulds also infect animal species—athlete's foot is a common example of this, so are ring-worm and scabies. Animal parasites include the tape-worm, the liver-fluke, and an amoeba responsible in tropical areas for dysentery. These parasites and related ones affect all sorts of animal life. Cancers too affect both the plant and animal kingdom.

The mode of transmission of disease and parasites about an ecosystem is complex. It may be through direct contact with a member of the same species, or through the atmosphere, or in a complex manner involving other organisms. Some tropical mosquitos for example carry the malarial parasite (a protozoan), which is injected into the victim's bloodstream while taking its meal. Disease may also be transferred through food organisms.

Although disease and parasitism have a trimming effect on populations by eliminating old, young, and non-resistant individuals, it is not to the advantage of a disease organism to wipe out the whole population, otherwise its further transmission might be terminated. In other words, the best-adapted parasites and disease organisms tend to be sub-lethal, and ones to which the host organism can acquire a partial immunity or resistance. Altogether disease is a complex and potent control mechanism.

3. PREDATION When the fox feeds upon the rabbit, or the wolf upon the deer, this is predation and it is a responsive population-control mechanism. Should for example, sunlight and rainfall be optimum and the resultant summer an excellent one, plant growth will be abundant, particularly new shoots and young saplings. Then, in response, the rabbit population will be encouraged and will increase its size. With the increased density of rabbits, the lynx during its hunting will come across many more rabbits than in previous summers. It therefore increases its predatory activity upon the rabbit population. But if the initial surge given to the rabbit population by the good weather cannot be reduced by the predatory action of the lynx then, winter permitting, it may increase still further the following summer. Because of the adequate rabbit catch, the lynx population may begin to echo the trends in the rabbit population, also increasing in size. Finally, predation by the lynx will push the rabbit population downwards. The excess lynx population now faces a food shortage, some of the individuals starve, and reproductive success is lowered. In this case, it could be argued that the rabbits, through availability of food, were controlling the lynx population, just as the lynx population, through predation, was controlling the size of the rabbit population. Such a dilemma is a difficult one to resolve;

perhaps it is simpler to say that both controls are at work at the same time. Lynx and snowshoe-hare populations have been observed to fluctuate in this manner.

Trapping records for these two organisms were obtained from the Hudson's Bay Company and plotted against time. The population of hare rises and falls in ten-year cycles for the whole eighty-year period covered. In a delayed response which echoed the hare cycles, the lynx population also rose and fell. The regularity of these cycles has been highly intriguing to ecologists who have yet to fully understand them, even though predation is thought to be an important factor.

Although predation is known to be an essential mechanism within ecosystems, its full status is very difficult to predict and very poorly understood, especially in conjunction with other mechanisms.

4. COMPETITION Parasitism, disease, and climate appear to act on all trophic levels of the ecosystem. Predation acts from one level to another. Competition appears to act between the organisms on the same trophic level. Competition can occur between populations of different organisms (interspecific) or between members of one species (intraspecific).

The wood-mice of our forest ecosystem are all engaged in a search for food material—seeds, nuts, and shoots. The most successful organisms tend to get the most food, and these organisms tend to be the healthiest and best adapted to the prevalent conditions. If we were to examine all the mice in the forest and grade them on their food-gathering activity, we would find a whole range of organisms, some good, some average, and others poor. Characteristically, the largest group of organisms is that with the average range of ability. In the competition for food, it will be the best adapted under the prevalent conditions that get the food, and the harsher the conditions the more and more selective competition will be for the most able organisms.

The wood-mice are not alone in their demand for nuts and seeds. Many small birds like the siskin and the waxwing also exploit these materials. Some of their food sources may be generally unavailable to the wood-mouse; for example some food is attached to cedar boughs. Others, however, that scattered on the forest floor or on the snow surface, for instance, may be commonly shared. Here then, competition occurs over the common food source. It will affect the population size of each species, depending on the individual success of each population at exploiting this food source, but also on the degree to which these organisms share this common supply. Again, with competition as with predation, although it may be fairly easy to assess the qualitative aspects of this control, it has proved extremely difficult to assess its quantitative impact, particularly in comparison with the other mechanisms at work in an ecosystem.

5. *TERRITORIALITY* Although this mechanism is at work within the population of a single species, it does not take place among all species. It is most important among populations of top predators like the owl, the wolf, and the hawk. It also works at other trophic levels, however. The fox will provide us with an example. The male adult fox will stake out a territory of its own. The dimensions of this territory may include physical features such as a cliff or a river; in other cases they may be less easily defined. The net result is one male fox for a certain area of ecosystem, an area that is capable of providing the fox and his family with enough energy for survival. This area the fox defends against all other foxes. Defensive activity includes menacing sounds and motions directed towards a rival in a manner necessary to assert dominance. This defensive behaviour rarely results in bodily damage. The only other foxes excused from this assertion of dominance and allowed within the territory are his mate and their cubs.

Much of the forest then will be staked out into fox territories; foxes without territory will be forced to the edges and to borderline areas. Only foxes that are landed, or tied to such a defined territory, will gain a mate and reproduce. Consequently, the generation of new foxes per unit area of ecosystem is limited.

Wolf populations appear to function in this sort of way, although territory is defended not in pairs but in packs. On the savannas of Central Africa, lions form packs called 'prides', which consist of one male, several females, and their young. These prides take their territories with them as they wander, but they do it in a way that leaves no doubt as to possession. The phenomenon of territorial behaviour encompasses intraspecific competition in the way that it works.

An interesting situation in which territoriality probably plays an important role is that on Isle Royale in Lake Superior. Studies have shown that three of the components of that ecosystem—wolves, moose, and vegetation are beautifully stabilized. Each moose consumes over five tons of vegetation annually, and if the moose population were large enough, it could be a seriously destructive force on the foliage. However, the population appears stabilized at a level just below the maximum sustainable by the vegetation. The wolf population, which plays a role in this, is stabilized at about one wolf per thirty moose, enough to support the wolf population but not enough to endanger the moose. Evidently control mechanisms are at work at both the level of the wolf and the moose.

Other types of organisms exhibiting territorial behaviour include certain monkeys, certain rodents such as the squirrel, certain ungulates such as the zebra, and many bird populations including robins, warblers, crows, and ducks, and fish that live near the margins of lakes such as bass and sunfish. Many bird songs are simply the male bird asserting his dominance over his territory, whether it be a bush or a tree, a field, or a whole section of forest.

6. AVAILABILITY OF FOOD For many organisms in the forest ecosystem this mechanism is a way of life. For many others it is the last resort of all the control mechanisms. The populations of the decomposer soil microbes depend very much on their food sources. Let us look at a bacterium that depends solely on chitin for its food source. (This microbe is an example of the highly specialized roles that the decomposers play.) When a beetle dies and has been pulled apart by the larger decomposer organisms, hunks of chitin-rich matter are released to the soil. This microbe then begins its work and its population expands in the presence of plentiful food-stuffs. However, the amount of chitin in that localized pocket of soil is limited to what was in the body of the beetle, and it is eventually depleted. At this point the microbe population collapses. Some die out, while others go into an inactive and relatively inert spore form. They remain in this state of dormancy until more chitin arrives on the scene. Again the population explodes, and collapses, and so on. (See Figure 15—2.)

This is not the sort of life-style one would predict for organisms such as the deer, the fox, or the bluejay. These organisms have neither a resistant dormant form like the spores of yeasts and moulds and bacteria; nor do they have the short and fast reproductive cycles of many insects that would enable them to replenish their population when conditions improved. Indeed, for the population of an organism to fluctuate between mass starvation from overpopulation and the dangers of extinction from underpopulation would be a disastrous situation. Should all of the larger components of an ecosystem function in this way it would be disastrous not only for the individual

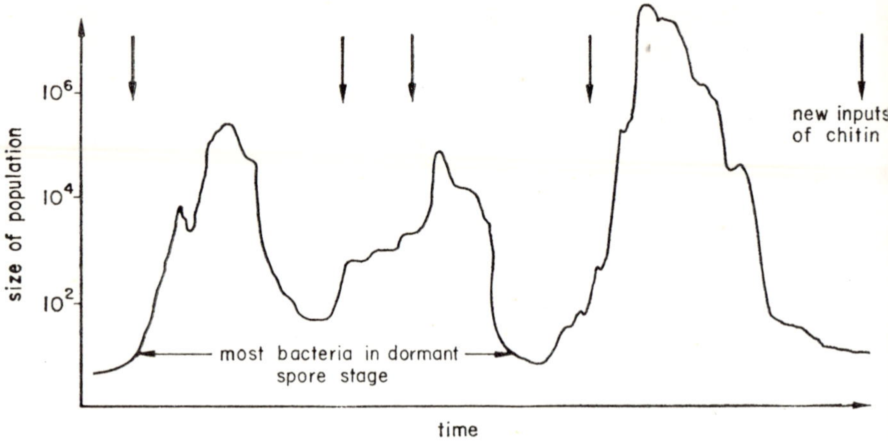

Figure 15—2 Fluctuations in a Soil Microbe Population

species but for the whole ecosystem. Thus, to prevent this situation, all of the control mechanisms acting together damp the sharp oscillations that often characterize natural populations of micro-organisms such as fungi, yeast, and bacteria. They have smoothed out the curve so that there is usually adequate food for the existing population. Each time one population attempts to rise and to reach a point at which it would run out of food on a widespread basis, disease, climate, territory, and predation pull it down. In other words, the point of starvation is never reached for a majority of the population because other controls prevent it from reaching this point. The more controls that act, and the more interspecific relationships that occur, the smoother the population curves appear, and this accounts for the great stability of an ecosystem.

These points are clearly illustrated by the consequences of human intervention in the dynamics of a deer population on the Kaibab Plateau in Arizona. Before 1907 the deer population was regulated at about 4,000 individuals by the predator populations of wolves and pumas. Then, between 1907 and 1923, a government programme was set in motion to exterminate the predators. By 1925 the deer population had exploded to 100,000 individuals and the plateau was heavily overgrazed, from grass to underbrush to leaves. In the following two winters 40 per cent of the herd starved to death, and the population continued to drop in size until it was below 10,000. Vegetative damage precluded a recovery of the herd even at this point and was evident for a long time after. It has been estimated that the maximum size of the deer population sustainable by vegetation was about 30,000 individuals, and the predators had been keeping the deer at a level well below this. Their removal forced limitation of food availability to come into play as a factor of control. At this point the deer population behaved in a manner analagous to the bacterial population.

An ecosystem, then, appears to embrace more than one community of organisms arranged about another, like aphids on a rose bush. It encompasses the concept of biological stability. In other words, an ecosystem consists of a community of different kinds of organisms interacting with each other in a manner that is conducive to the stability of the whole community. Through these complex interactions based on the availability of energy, and consisting of parasitism, disease, predation and territoriality, the ecosystem (conditions of climate and geology permitting) is a self-perpetuating unit, free from random and sharp oscillations that would ultimately lead to its downfall.

Despite the inherent quality of stability within an ecosystem, it is still not an easy concept to define. The degree of stability of ecosystems appears to vary, and one can indeed be faced with the question: at what point does the stability of a community merit it being called an ecosystem? Again the answer is still forthcoming, and it can only be said that the concept of an ecosystem is a valuable one, even though we are not able to completely delineate its limits. Let us further explore the concept of stability.

Stability and Diversity

A fundamental limit on the *biomass* of an ecosystem (the weight of all living organisms in that system) is the availability of energy. The biomass at every trophic level ultimately depends upon how much energy is fixed during photosynthesis by the green plants. But how this biomass is distributed among different types of organisms is a different matter. If in one of two ecosystems having the same biomass, there are a greater number of species, this system is said to be more *diverse*, while the other is said to be less diverse. Diversity in some ways appears related to the opportunities offered to living organisms by an ecosystem in the way of ecological niches, or ecological 'professions'.

Diversity appears in some ways related to the number of opportunities for existence offered living organisms by an ecosysem in the way of ecological niches or professions. It appears that diversity has arisen, at least to some extent, in response to competition for available food sources, with increased diversity resulting in the broadened range of exploitation of food sources. Thus we find seed-eating birds of many types. There are those feeding on herbaceous plant seeds like the burdock, those feeding on the seeds of conifers like the cedar, the jackpine and the tamarack (with different species often feeding in different vertical levels of these trees), those feeding on the seeds of hardwoods like the maple, those feeding on the seeds in the litter of the forest floor, and so on. In every ecosystem these fine divisions between niches can occur. It could be noted that these divisions often occur temporally (in time); for example two different organisms may exploit the same food source but at different times. The hawk and the owl both exploit the mouse population in a meadow ecosystem, but the former during the day and the latter at night.

The number of 'opportunities' for various life-styles that are offered by an ecosystem appear to be related to the local conditions of climate and geology about an ecosystem. The most important factor in producing diversity in an ecosystem appears to be the productivity of that ecosystem—the rate at which the producer organisms fix carbon dioxide and supply energy to the rest of the system. The conditions of temperature in an ecosystem appear to be one of the most important factors in controlling this rate. The higher the temperature, the higher the metabolic rate of the cellular processes within an organism (at least below a certain upper limit of temperature) and thus the faster the rates of turnover within the ecosystem. The most diverse ecosystems appear to be tropical ones, such as that about a coral reef. Equatorial rainforests reflect this high diversity in the great variety of plant forms and adaptations and the hundreds of thousands of bird and insect species all occupying niches—some of them very finely divided—about these producers. Movement of matter between all of these niches through the food webs is very rapid in tropical regions.

In the arctic, the metabolism within an ecosystem is much slower, consequently there appears to be less diversity. From the producers through to the top predators there are fewer varieties of organisms at each trophic level, and these are characterized by species that are well-adapted to a long cold winter and a short fast-growing season, where productivity is limited to the organic material fixed by small shrubs, flowering plants, grasses and lichens. Between the tropics and the polar regions there appear a spectra of ecosystems at varying stages of diversity. At any one point between the arctic and the tropics it is difficult to define this transition, but it does seem to occur.

The degree of diversity of an ecosystem seems to be related to its degree of inherent stability. By this we mean its ability to withstand fluctuations in its component populations, and small changes in weather and productivity that in a seasonal sense the ecosystem has had the opportunity to adapt to for many thousands of years. The large numbers of organisms making up the food webs of tropical ecosystems contribute to a greater smoothing of the population oscillations of any one species and allow the system to adjust itself to changes in its component parts and in the indigenous conditions. On the other hand, arctic ecosystems appear to be characterized by a lesser stability that can be noted in more violent fluctuations of population. Temperate forests and grasslands are somewhat between these two extremes. In another way, this inherent stability of an ecosystem could be termed its *persistence*—its ability to maintain itself under a specified range of conditions. That the persistence of an ecosystem is directly related to the diversity of that ecosystem could be termed the *Third Law of Ecology*. When one considers that each ecosystem is a functioning unit of the biogeochemical network of nutrient flow, the importance of ecosystem persistence becomes important: in the long term persistence on an ecosystem level is contributing to the stability of the biosphere as a whole.

Stability however can take on a meaning other than that of persistence, The ability of an ecosystem to withstand forces to which it has had little or no previous experience could be termed its *resilience*. These are forces that are too sudden and sporadic, too violent, or too rare to allow it the chance of adaptation. They include climatic forces such as floods, geological forces such as volcanoes, landslides and fire, as well as many of the forces of human technology. The stability of an ecosystem under these pressures, unlike persistence, is not directly linked to its diversity, in fact in some cases the contrary may be so. Re-establishing a biological balance in the wake of forces such as fire and advancing lava requires a very different approach than that in an ecosystem where productivity has been particularly low for a summer. The mechanisms that are employed by the biological world in these instances are termed community succession. The speed with with an ecosystem re-asserts itself through these mechanisms is a measure of its resilience.

Biological Community Succession

In a given locale, the factors that initially affect a community of organisms—and to a large degree are determining—are the geological nature of the substrata and the conditions of climate suspended over that area. These factors tend to be uncontrollable. Individual factors like the mineral content of the rock, the degree of erosion, the ambient air temperature, amount of sunlight etc., are important. Ideally, one is considering the situation much as it existed before the first simple life emerged from the oceans. However in more modern times it is often the situation following a landslide, a fire or a volcano.

The settling of an area by living organisms usually follows a somewhat reproducible pattern or a succession. Succession parallels, in some ways, the first invasion of the land masses by the green plants, following the oxygen revolution. The first species to invade successfully are those tolerant of the conditions imposed by the climate and by the geological substrate and are organisms like lichens and mosses. These germinate and give rise to the first life-support system. Given adequate nutrient or rainfall, this first life will continue to transform the local environment, by modifying the basal rock into smaller particles and by laying down organic material. To a certain extent this community will be able to support very small herbivores such as mites. But as a result of the transformation, this environment becomes more suited to other organisms, and a new wave begins to invade. The richer soils are more suited to small grasses which replace the first plants. These in time, further transform the environment: richer soils accumulate, so do decomposer bacteria and nitrogen fixers. Roots begin to regulate the flow of ground water, and plant cover protects the soils from direct rainfall, wind and sunlight. Eventually this community may be succeeded by another, then another, and so on. Eventually a community more persistent than all the rest will arise; this is termed the climax community. It is one in which living organisms have exerted the maximum possible control over the local geological and weather conditions. Figure 15—3 illustrates this succession as bare rock is converted over hundreds of years to forest.

The succession of plant communities does follow a reproducible series of events. There is however a certain degree of variability in-

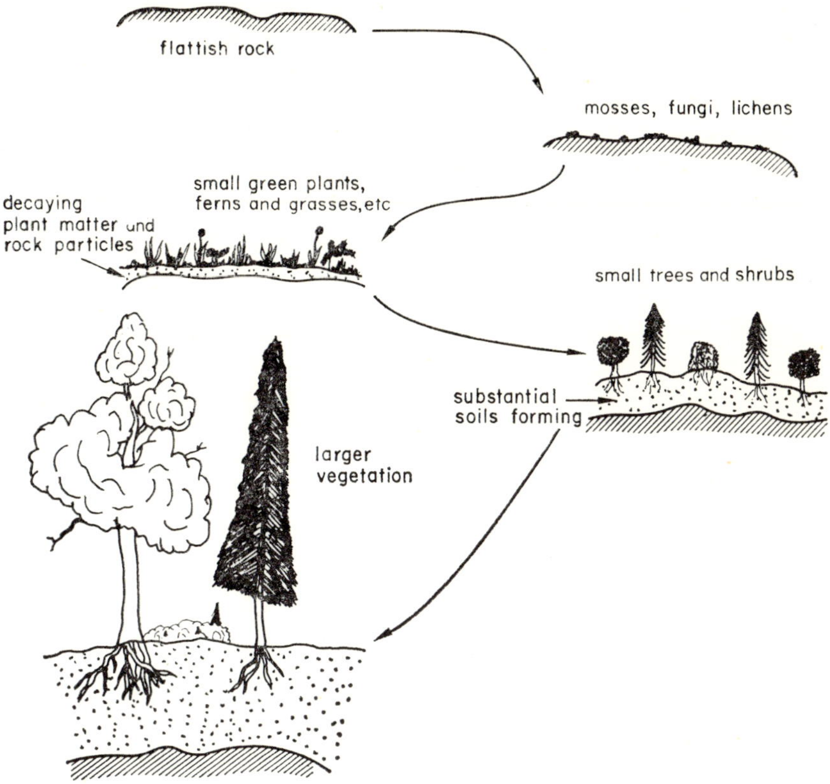

Figure 15—3 A Plant Community Succession

volved. The invading species will depend on the movement of seeds and seedlings from surrounding areas, and these may vary greatly from one part of the Earth to another. In fact several organisms may compete for one phase of a succession, and their individual success may be dependent upon local geological or climatic peculiarities. And again, the next phase of the succession will be at least partially, if not totally, dependent upon the preceding community.

Accompanying changes in plant communities are changes in animal communities and ecosystems. At any point in time these will be balanced, stable systems (although the persistence will be related to the diversity of the successional stage), but in the long term these systems will be slowly changing. These communities change with many factors. Bird species will change, for example, with the appearance or disappearance of under-brush, insect and plant food species, nesting places and ground climate. For instance, as a community changes from a grassland to a hardwood forest, noticeable changes in bird type occur. Grassland species like the meadowlark and the grasshopper sparrow are replaced by the species inhabiting shrubs and low trees like the cardinal and the yellowthroat. As conifers begin to predominate so do birds like the nuthatch and the vireo, only to be replaced in the climax oak-hickory forest by the characteristic woodthrush, woodpeckers and warblers.

However, the transitions which occur with succession are not abrupt and sudden; they are slow and continuously occurring phenomena, in some cases taking hundreds of years. But they occur, and the feeling of many biologists is that the climax is a condition towards which all plant communities proceed. They may not always be reached, perhaps due to climatic or geological limitations, and sometimes even due to organisms within them. It was noted in the south of England that the downs (rolling hilly areas) were perpetually at one level of succession, and that the next step which had not appeared involved the growth of small shrubs and trees. There are two major mammalian herbivores on the downs, the sheep and the rabbit. When myxomatosis hit the rabbit population one year, there were visible signs that the plant community was pushing towards the next stage of succession. In other words, the combined grazing action of both mammals held the community at a non-climactic stage.

It appears too, that climax communities are the most diverse of all of their predecessors, in other words they are probably the most persistent. We have already seen how climax communities tend to maximise biotic control of the abiotic world. The greater diversity and stability of climax communities is reflected in another ecological measure, the P/R ratio. This means that one places a value for the total productivity of the system (as measured by green plant production per unit area per year) over a value for the total respiration that has been carried out (the amount of the productivity that has been burned up for energy needs). What has not been burnt up has of course been incorporated into the bio-mass—tree trunks, hair, topsoil, etc. Thus as long as the system is growing P/R should be greater than 1, or, more organic matter is created than used. When growth has stopped, P/R should be equal to 1, because on the average no net growth is occurring. And when P/R is less than 1, then the system is dying, and living matter is being consumed for energy faster than it can be replaced.

Pre-climax communities appear to be characterized by a P/R ratio than is greater than 1, climax communities by one that is equal to 1, and declining or polluted communities by a ratio of less than 1. Perhaps the *Fourth Law of Ecology* could be formulated about the concept of climax. It might be stated: all ecological communities strive towards a climax community, that under the indigenous climatic and geological limitations, maximizes biological control of the community environment, and maximizes persistence through diversity.

Following any forces to which a biological system is unable to adapt, the process of community succession begins again. Dependent upon the local conditions (of nutrient supply, rainfall and sunlight), the severity with which the force destroyed the preceding community, and the state of diversity in the original community, the process of succession will occur slowly or rapidly. Although some of the most diverse and persistent ecosystems on earth, tropical rainforests have a low resilience towards forces with which they have had little experience and are slow at re-establishing themselves. Part of the reason for this lies in the great severity of tropical conditions—high temperatures and heavy rainfalls that rapidly leach nutrients from the bare soils. The great productivity of these ecosystems rests in the very efficient recycling mechanisms that have evolved to retain nutrient within the system. Simultaneously, the great growth of plants gives protection to the soils and modulates the climatic conditions beneath the jungle canopy. Removing the rainforests removes all of these mechanisms and succession must therefore fight against harsh conditions to re-establish a similar climax community.

Temperate ecosystems appear to regenerate most readily under the pressure of these external forces. In these regions of moderate sunlight and gentle rainfall, plant communities can quickly re-establish control following a fire. In the arctic and sub-arctic areas of Earth, resilience is again similar to that in tropical regions. Low temperatures and long periods of winter create conditions over which it is already difficult enough for living communities to exert control. Destroying the delicate balance between the living communities, the soils and the permafrost can create wounds in the tundra that take decades to heal. In parts of the Canadian arctic, vehicle tracks across the tundra can still be seen twenty years after they were first created!

Evolution and Change within the Biosphere

Events which alter the environment in too sudden a manner may destroy much of the life within it. But over the course of millions of years, these changes can lead to new varieties of life and stabler ecological conditions. And it is within this whole background of change that evolution and environments must be viewed.

Outside the biosphere, things are to a remarkable degree very constant; the seasons, the years, and the amount of sunlight continue to flow in an almost constant manner. Sunlight itself may vary from time to time with sudden short solar outbursts, or with slow heatings or coolings. Some of these may have profound long-term effects on the climate of Earth, and consequently could be potent forces in evolution.

Conditions on the planet are not nearly as consistent as they are in the solar system. Geological activity is constantly shaping the surface. About 500 million years ago Earth is thought to have been inhabited by a single continent, Pangaea. From then until the present, Pangaea has split many-fold, its subcontinents carried by forces within the earth's crust to their present positions. Today this process has not ceased, and continues to transform Earth. The distances that these continental masses move yearly are relatively small, and almost unnoticeable. But in the evolution which occurred in the half a billion years that the continents moved apart, these must have been extremely powerful forces indeed. Populations were isolated, and brought together again. Climates changed as the continents moved through different climatic zones on Earth.

With Continental Drift, other geological processes have been going on: mountain building, and island building, for example. Land areas have sunk beneath the waves, and land-bridges have been formed. River valleys have been created and eroded, flood plains formed, and alluvial deposits extended the edges of the continents into the oceans. And glaciation in the last twenty million years appears to have swept back and forth across the northern hemisphere four times.

But none of this occurred quickly, and thus life could adjust, producing lines of organisms that were more suited to the environmental changes. On the other hand, certain geological changes occurred within periods of days to hundreds of years. Fires swept across grasslands as a result of lightning, eroded rock particles caused silting up of river valleys, landslides devastated parts of forests. For these changes life turned not to selection but to succession as a means of adjustment. Plant and animal species reinhabited the area in an ordered manner to reassert the maximum of biological control. Over the lifetime of a single individual, the environment remains fairly consistent.

Life as a whole interacts with the environment in three major ways. Firstly, it transforms it as much as possible, to a hospitable state. On a global scale, it has produced the biosphere with its protective layer of ozone and the regulatory and recycling mechanisms within it. As a geological force, life has also altered the environment to its benefit, by slowing down erosion and forming stable soil systems, and by creating extensive reefs in warm shallow seas.

Secondly, life adapts to the forces which it is unable to change or to the extent that it is unable to change them. Adaptation has occurred over the long term, either to short interval and regularly repeating changes, such as the tides and seasons, or to slow unpredictable changes. Take for instance the oyster, which when moved hundreds of miles from the seacoast to an aquarium, still opens and closes its shell to the pattern of the incoming and outgoing tides! To short, sudden changes that it is unable to control—changes like volcanic eruptions—life replies with succession.

Thirdly, life adapts to changes that it has created in the environment. Through millions of years of evolution, plants, animals and bacteria have evolved slowly to the changes wrought by each other on the environment. Subtle patterns and interrelationships occur among all levels of the food webs and between food webs in many ecosystems. Again, this has been a process of slow self-adjustment to life's own changing conditions—again producing a more stable situation.

Think, then, of evolution as an arrow, leading from a chaotic and extreme environment to one of order and moderation. Along this arrow, organisms appear and disappear, but on the overall scale a diversification has taken place, accompanied by great selection. But think of a cross-section of the arrow taken at various points along time. At each of these points one would see the development of a stable biosphere as single ecosystems become meshed together into a total global system, and each point would appear suspended in a state of balance.

CHAPTER 1—6

The Human Animal: Phase One

And what rough beast, his hour come round at last
slouches toward Bethlehem to be born.

W. B. Yeats

Up to this point in the book, we have ignored the presence of human organisms in the biosphere, primarily in an effort to understand the fundamental ways in which living systems operate. The human context, like that of all other living things, began in the inseparable totality of this biosphere. Sometime in the geological era known as the late Pliocene, perhaps four million years ago, a line of apes evolved into human form. Later, through the use of technology, these organisms were to enter a new state of existence that was to have almost catastrophic effects upon their environment. Let us consider, however, this primitive human organism before the technological adventure, as a focal point of a late Pliocene–early Pleistocene ecosystem. Perhaps in this manner we shall be able to elucidate some of the dependencies between humans and their environment.

Establishing a transition point between primitive and technological humanity is sure to arouse controversy. Many arguments can be presented in favour of use of tools, speech, agriculture, as markers of this transition point. However, biologically speaking, one can argue that the use of fire represented the turning point. Although the use of tools preceded the use of fire, this development was not, and is not, unique in the biological world. In the course of evolution many other species have learned the use of tools. One type of African bird, for example, drops large stones from above in order to break open ostrich eggs on which it feeds. One of the famous finches observed by Charles Darwin in the Galapagos holds a thorn in its beak to pry grubs and other food insects out of tree cracks. And many forms of monkeys use sticks and rocks in the motions of defence. However, only humans have developed the active use of fire as both a protective and a hunting agent, and one that we shall later see also had widespread environmental effects. 'Primitive' will therefore be defined as the stage of human life before the advent of fire-use.

In previous chapters we explored the processes by which various lines or species of organisms arose through evolution from one type of organism through the course of millions of years. We saw the spread of life onto the land and the filling of many ecological niches, through the processes of diversification and speciation. By sixty

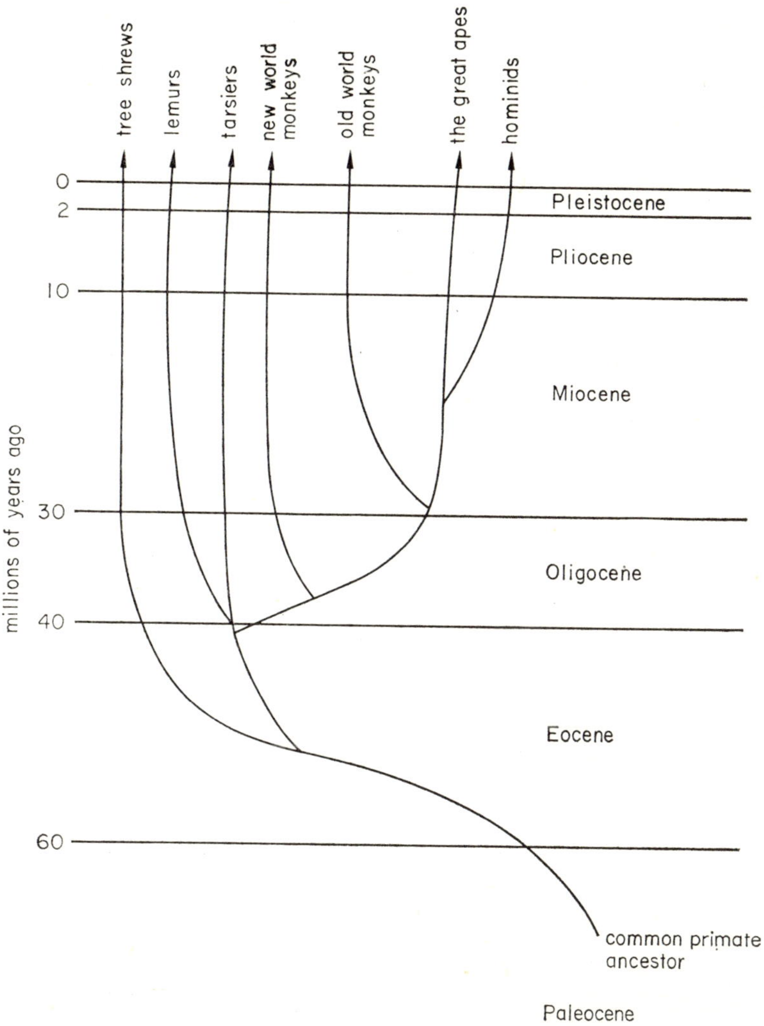

Figure 16—1 Diversification of the Primates

million years ago, at the end of the Paleocene era, most of the major lines of mammals had evolved. From then to present times each of these mammalian lines, including that of the Primates, has further diversified. (See Figure 16—1 for an outline of this.) At the end of

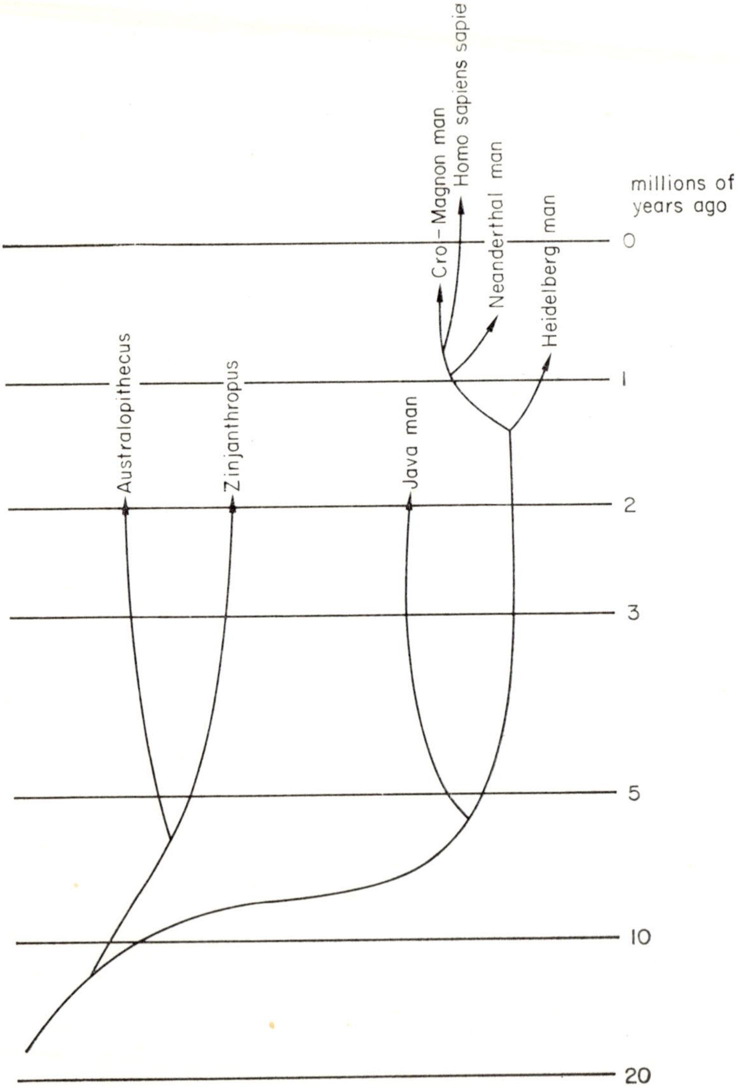

Figure 16—2 Evolution of the Hominids

one of the primate lines is the human race.

At many points these lines of diversification ended in extinction and no descendants remain today. Many lines still exist; for example, modern surviving lines of the great apes include the gibbons and the gorillas. Like each of the other lines, that of the Hominidae also diverged. Figure 16—2 outlines the general pattern of divergence as it is speculated today. These divergences began less than fifteen million years ago, but of all the resultant species the human is the only one still surviving.

What was the environmental stimulus that selected the hominids from the old-world monkeys and the great apes? Present evidence seems to point to the advance and retreat of the glaciers as the factor. It seems that some twenty million years ago, eighteen million years before the Pleistocene began, the climate on Earth was changing. Not a fast change but a slow one, taking some eighteen million years to reach its climax. Ice began increasing in the polar regions of Earth. The reasons for this are not yet clearly understood, but as a result more and more water became tied up in the form of ice, and in the tropical areas of the planet temperatures decreased and rainfall decreased. Subsequently, the rich, lush jungles shrank to patches in an increasing sea of dry savanna grassland. In the driest areas deserts expanded. This slow transition went on until the beginning of the Pleistocene. Then suddenly the first glaciers swept down over the northern and southern hemispheres. Four times in a million years they advanced and retreated. The last retreat began only 50,000 years ago.

Most of our primate relatives live in the tropical regions of earth, and many are jungle-dwellers. Although a few of them are insectivores, most are herbivores, feeding on vegetable matter such as plant shoots, tubers, and fruits. In addition most are tree-dwellers, or at least versatile in trees, both for feeding and defensive purposes. As the late Miocene jungles decreased in size, intense competition for food forced some of the primates out onto the savanna. Here, away from trees, they were defenceless and face to face with predators such as the lion, against whom they had little chance. Except for a few tubers and fruits they were away from major food sources, and as hunters they lacked the good eyesight, speed, and tracking abilities of the top predators. Adaptation to these pressures is thought to have produced the hominids, and many anthropologists feel selection was based on steadily increasing brain size. Banded together, a reflection of their monkey past, these primates perhaps started as scavengers, sharing with the hyena the remains of the lions' feast. Eventually they acquired the use of tools and became primitive hunters. Consequently they left their ecological role as herbivores and as prey, taking on a new role as top predators.

Much of the recent investigation into the origin of the human species has taken place in Central Africa, particularly at Olduvai Gorge. It has become evident, through work done here by Dr L. S. B. Leakey and his colleagues, that early hominids evolved on the African continent, and that during the late Pliocene–early Pleistocene period there were probably several distinct species alive. One of these lines was *Australopithecus*. About ten million years ago *Australopithecus* was beginning to speciate from the rest of the hominids (refer to Figure 16—2). Already erect to a far greater extent than his ancestors, *Australopithecus* was, in appearance, somewhat like the human of today, and a user of tools. A diverse eater, the remains of many organisms from large mammals to birds have been found about his habitations, which were usually caves. Many of his prey exhibit

Figure 16—3 Australopithecus and its Food Web

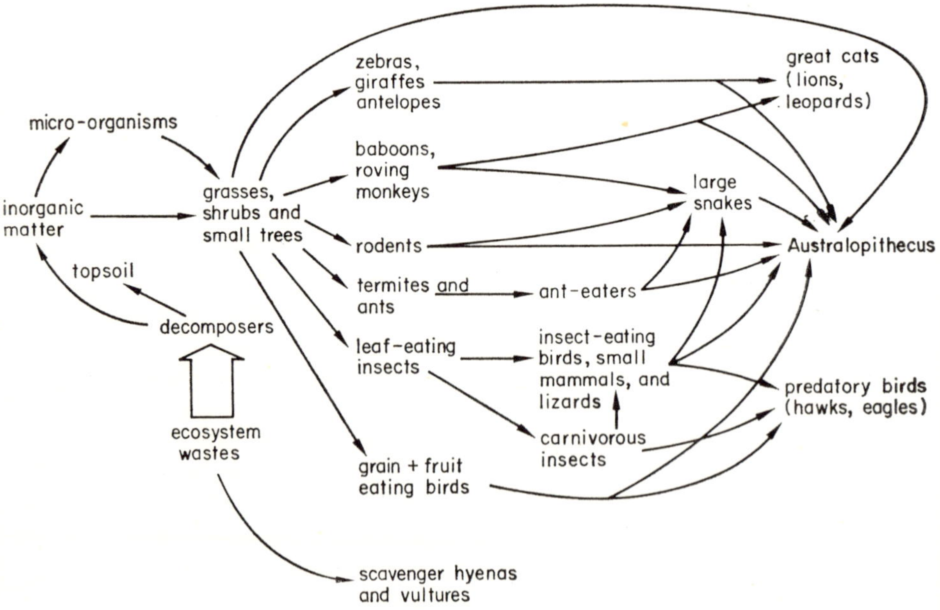

signs of sudden death by a sharp blow applied to the head area, evidence of the use of a bone or stone as a weapon.

Physiologically and biochemically there were differences between early *Australopithecus* and the human of today, but because evolutionary changes are slow, and relatively speaking, time between the two is short, there are many striking homologies. We are, then, talking about an organism closer to us than any other in the biosphere today. We are also, because the use of fire did not occur until some nine million years later, looking at what was probably a pre-technological human, one, therefore, functioning in a traditional biological sense as part of its ecosystem.

How did *Australopithecus* fit into his ecosystem? A speculative place is shown in Figure 16—3. The Pliocene fauna was probably different from that of today, although many of the organisms would show some sort of resemblance to their present-day descendants; some would have no descendants. But most would be in ecological niches similar to those occupied by their descendants today. Rather than reconstructing the whole Pliocene ecosystem, it is perhaps easier to paint the picture in terms of a present-day savanna ecosystem—one probably much like that of *Australopithecus*.

It is probable that *Australopithecus* was a territorial animal—simply a carry-over from the past, since nearly all primates are territorial. In addition he was probably a social animal living, like the monkeys, in groups that had a definite system of hierarchy. Both of these characteristics were advantageous in hunting and protection. It is most likely that bands or tribes of *Australopithecus,* like the other top predators, staked out areas of the savanna as their exclusive hunting domains. The weaker bands were forced to the less fruitful areas of the ecosystem and suffered, as a result, higher mortality. There are very interesting implications in this, because these are the sorts of behaviour exhibited by other top carnivores. In other words, the transition of our ancestor from prey to predator was possible because they already possessed instincts of territoriality, an important feature of population control at the tops of the food chains.

Another important control mechanism on the populations of our australopithecan ancestor was disease. Because mammals can become resistant to infection through the development of antibody systems, a certain resilience on the part of *Australopithecus* was possible and consequently it was mostly the old and weak that suffered. However, if a whole pocket or tribe was extinguished, the territoriality that divided the species was also effective in preventing the further spread of that disease. The controlling effects of climate, that is, extremes of temperature, probably acted in much the same way as disease—on the old and the weak.

It is possible that a slight population control was enacted through predation. For example, other top-niche carnivores like the lion and the leopard would occasionally take their toll, but this was probably equalled by a counter attack of *Australopithecus* on an occasional cat. Only in times of severe food shortage, when competition among the various groups of top predators became most intense, would inter-predation likely occur to any great degree. And this would have a regulatory effect on the ecosystem, cutting down the pressure on the lower trophic levels.

In total these mechanisms helped ensure that australopithecans did not outgrow their food sources, nor overgraze them. As top predators australopithecans spread their food chains widely. Probably retaining some vegetarian habits from earlier days, they expanded their diet to lizards, small mammals, ungulates such as the antelope and the zebra, and probably even grubs and other insects. Because the food chains were wide and branched, no organism felt the full brunt of their feeding habits, and as with the foxes in the temperate forest ecosystem, when one food source became low, they tended to concentrate on others.

For the availability of food on a continual basis our ancestors were not only dependent upon the population-control mechanisms in their own species, but also on those of every other organism in the bio-sphere. They were, for example, dependent on their fellow carnivore the lion. Although the two organisms shared some of the same prey, the emphasis was different. For example, the lion was more likely to feed on antelope than on birds such as the guinea hen and the other smaller species preyed upon by *Australopithecus*. Effective population control in their ecosystem therefore required their joint roles. Say, for example, that the lion population did grow out of control. Because our ancestors hunted some of the same prey, they would tend to shift emphasis as that prey became short. Not only would the prey of the lion become endangered, but so would the prey of our ancestors.

The lion of the Pliocene kept the grasslands under a control bene-ficial to our ancestors, just as they were interacting to the benefit of other organisms. It is important to recall here that efficiency of the food energy passed on from one trophic level to the next is about

10 per cent of that trophic level; and that each of the populations is neatly in equilibrium with a certain amount of flexibility to allow for annual fluctuations.

Nutritionally there were a series of dependencies stretching between australopithecans and the roots of their ecosystem. Primarily all of their energy came from the sun, and they were, therefore, reliant on the green plants to fix this energy. If they consumed fruits or tubers they were getting this directly. However, in eating meat they were receiving this energy indirectly and were thus dependent on the herbivores as well to convert it to animal tissues. In some cases this dependence involved even another converter if, for example, they consumed a large snake or lizard.

Some of this conversion is not so obvious as it apparently seems. Much of the energy stored in plant matter is unavailable directly to mammalian organisms. The reason is a chemical one: most of the energy-rich glucose produced by green plants is stored in the structural material of cellulose. To release the energy-rich subunits, this cellulose must be broken down, and mammals do not possess the enzymic ability to do this. Herbivores such as the antelope have overcome this problem. In one of a unique series of up to seven stomachs they maintain special bacterial cultures. These cultures, in return for sharing some of their hosts' food, secrete a cellulose-digesting enzyme; consequently these herbivores can eat grass and leaves, materials otherwise poor at meeting mammalian demands. Thus, as the herbivores themselves were, *Australopithecus* was dependent on the bacteria in the herbivores' stomachs for their ability to degrade cellulose.

The dependencies incurred by our ancestor in obtaining nitrogen can be explored in a number of different ways. Of the twenty amino acids found in the protein matter of living organisms, ten are essential to the human being and likely were also essential for *Australopithecus*. They are essential because the human body is unable to synthesize them in any useful amounts. In the proportion of these ten amino acids, plant matter is distinctively different from animal matter, and the latter is by far the best dietary source of these amino acids as far as animal needs go. Herbivores get around this problem by continual grazing and accumulation over long periods of time of much plant matter. Thus *Australopithecus* was dependent on the herbivores for the concentration and proportionation (in the correct balance of amino acids) of protein matter.

But protein nitrogen is only available in the first place because the green plants have fixed nitrate and ammonia from the soils, and this is only possible because the nitrogen-fixing bacteria have converted atmospheric nitrogen to ammonia, and the nitrifying bacteria have converted ammonia to nitrate. On the average for example, the bacteria in an acre of fertile topsoil fix seventy pounds of nitrogen annually. This can range up to 400 pounds per acre in a single season in tropical areas! Through similar pathways many other mineral essentials are brought into the food chains, whether through bacterial fixation (as sulfides are converted to plant-usable sulfate), or directly through plant roots as potassium, iron, magnesium, and copper ore. Again to the advantage of the australopithecan, the primary consumers concentrated these minerals for him by the consumption of great amounts of plant matter over long periods of time.

Just as there are amino acids that most mammals are unable to produce themselves and thus must get from the environment, so are there other essential dietary materials that must be obtained this way. Three of the four series of polyunsaturated fatty acids—linoleic, linolenic, and palmitic—are dietary essentials. The dependence lies on the green plants to produce these and on the herbivores (particularly the seed-eating ones, because these fats are found in highest amounts in seeds) to concentrate them. Other dependencies relate to the vitamins that are outlined in Table 16—1. Many of these reach

Table 16—1 Food-Chain Origins of Some Vitamins Needed by *Australopithecus*

Vitamin	Producers	Step to consumer
Vitamin K and biotin	Intestinal bacteria	Direct absorption
Vitamin C (ascorbic acid)	Green plants, animals except the primates	1° and 2° consumers
Thiamine and riboflavin	Green plants, particularly grains	1° and 2° consumers
Pantothenic acid	Green plants, to a small extent, and micro-organisms	1° and 2° consumers Green plants, 1° and 2° consumers
Vitamin B_{12}	micro-organisms	Green plants, 1° and 2° consumers

right down to the base of the food chains; for example, vitamin B_{12} is synthesized only by certain micro-organisms upon which all other living organisms are dependent, from the green plants through to the top carnivores.

For the air they breathed australopithecans were dependent upon all the regulatory effects of the biosphere as a whole—long-term production of oxygen, re-absorption of carbon dioxide by the green plants, and the fixation and release of nitrogen. These provided them with a buffered environment in which there was protection from ultra-violet radiation, in which the temperature range underwent only mild fluctuations (4-50°C—mild in a galactic sense), and constancy of atmospheric pressure.

Although *Australopithecus* drank fairly impure water, it was probably more wholesome than the water we drink today. The toxic materials that must be removed from today's water supply and the infectious organisms that we remove by chlorination were likely present only at low levels at the time of our ancestors. The waste-recycling mechanisms of their ecosystem were geared to handling the population load of the ecosystem, and such wastes were probably well dispersed over the whole system.

Our ancestors' wastes included a large range of materials: scraps of unconsumed prey; fecal wastes in the form of semi-digested food matter and bacterial flora; urinary wastes including dissolved urea and ammonia. In the urine and feces were many compounds of nitrogen (ammonia, skatole, indole, and cadaverine) and sulfur (mercaptans and hydrogen sulfide). Many of these were toxic to *Australopithecus*, which is partially the reason they were secreted. Some smelly wastes may have been produced partly for warning to the organism to stay away from these wastes, dangerous because of their high bacterial content. This required an appropriate co-evolution of smell senses.

The process of decomposition began within the intestines themselves, through the action of their bacterial populations. Once the wastes were eliminated, the process was continued and completed by the scavenger maggots and beetles, and by the protozoa, yeasts, moulds, and bacteria of the soils. Ammonia was converted by certain bacteria to nitrogen and nitrate. The more complex molecules excreted served as energy sources for other specialized organisms, until most was converted to carbon dioxide and other environmental constituents. It has been estimated that the top six inches of fertile soil contain more than two *tons* of fungi and bacteria per acre, equivalent metabolically to some tens of thousands of people!

Australopithecus was dependent on this tremendously powerful activity for the elimination not only of personal wastes but also of those produced by the other top predators, herbivores, carnivores, scavengers, and green plants. This same activity was turning these wastes into fertile topsoil, stimulating new plant growth, and encouraging the food chains in which the pre-humans lived.

This was the very balanced and harmonious world of our Pliocene ancestors: a world where they functioned as part of nature, in which the end of life was the beginning, and their give and take as top predators contributed to the stability and preservation of the world in which they lived. Some time after the beginning of the Pleistocene, *Australopithecus* became extinct. Surviving it was another line of ape-people, perhaps a little more intelligent, for the advent of the Pleistocene glaciers meant a rapid climatic change, geologically speaking, and probably even greater selective pressure than that which had produced the first hominids. But this new breed knew the use of fire; they had forgotten innocence and were moving forwards into times of hope and despair.

PART TWO

Leaving the Garden

CHAPTER 2—1

The Human Animal : Phase Two

Environmental problems, far from beginning in recent times, had their start with the first human use of technology. Before the advent of agriculture, an estimated thirty billion fire- and tool-using people had walked the earth. Although at any one time there were far fewer people alive than there are today, the incessant use of these early technologies over thousands of years had profound effects upon global ecology. The first environmental changes probably occurred so slowly that human memory began to lose track of them. Early agriculture, the overgrazing of livestock, and the cutting of forests may have accelerated soil erosion. In areas such as northern Africa the land may have been transformed to desert. Yet within several generations people had moved on and had forgotten that anything but desert land ever existed there.

Technology, it can be argued, began with fire and sophisticated tools. Agricultural communities based first upon domesticated animals and later upon the raising of crops followed. Then the first cities arose and with them metallurgy and transportation networks. All of these had effects upon the environment, either upon plant and animal communities, upon the soils, upon the balance of ecosystems, or upon the population-control mechanisms within the human population. In this chapter we will explore the origin of technology and some of its first environmental impacts.

Fire and Weapons

About a million years ago on the grasslands of Central Africa, our ancestors first achieved the use of fire. It may have served as a protection from predators and climate, but later it became a hunting tool and was used to drive game into traps where it could easily be killed. Later it became a cooking tool; this eliminated some of the parasites and disease organisms that were present in food.

Although simple tools have been discovered that date well before the first use of fire, these consist of little more than crudely shaped bones, rocks, or sticks, and, as discussed earlier, these may have been

Table 21—1 A Recent History of Human Evolution

Years ago	Period	Human Population Size	Technology or Event	Effect on the Biosphere
1,000,000	Lower Paleolithic	125,000	Fire-use (Africa) Tool-use	Increased incidence of fire
300,000	Middle Paleolithic	1,000,000	Migrations to Europe & Asia	Pleistocene over-kills in Europe & Asia
25,000	Upper Paleolithic	3,300,000	Migrations to N. & S. America Domestication of the dog (?)	Pleistocene over-kills in N. & S. America
10,000	Mesolithic	5,000,000	First farming in the 'Fertile Crescent' Domestication of sheep, goats & cattle	Beginnings of soil erosion in the Middle East which spread with farming to North Africa & Europe
8,000	Neolithic	86,500,000	Early villages and small communities, beginnings of many crafts Use of gold	Accelerated soil erosion
5,000	'Civilization'		Use of copper, first city states in Mesopotamia Beginnings of trade	Transfer of plants, animals & disease about the Eurasian world
3,000	Iron Age	133,000,000	Use of iron	
1,971 (0 A.D.)			Roman, Indian, & Chinese civilizations	
1,000	Middle Ages	200,000,000	Maori settlement of New Zealand	Pleistocene over-kill in New Zealand
600		300,000,000	Black Death sweeps Europe	Extensive deforestation in Europe
300		550,000,000	Colonization of 'New World' by Europeans	
150		900,000,000	Beginnings of modern medicine	

no more than a normal biological phenomenon. More advanced tools, however, such as highly carved and polished rock blades, some of these fastened to wooden handles, date from less than a million years ago. Their use, like that of fire, appears to have evolved in the savannas of Central Africa, and their employment must have conferred a superior hunting ability upon their users.

The use of tools and fire stimulated the first increase in human population size. In a period of 700,000 years, the human population rose from about a hundred thousand to about a million. As the effectiveness of tools increased, in the following 300,000 years it increased to five million (see Table 21—1). The human race fast approached the limits of supporting itself in the ecosystems of Africa, and human migration to other parts of the globe began. Particularly in colder regions, fire became a distinct advantage. The use of fire and tools, however, was not restricted, in effect, to human population alone. Their forces were felt in what has been termed 'the Pleistocene over-kill'.

The Pleistocene Over-kill

The movement of the Pleistocene glaciers back and forth across the northern hemisphere had strong selective effects upon the mammals and birds of the preceding Pliocene era. A wide variety of new species appeared, many of them very large. They included giant condors, giant camels, mastodons, and the Great Irish Deer. All of them managed well during the glaciations, and as the last ice-sheets retreated northwards many of them were still extant. Yet today only Africa has a Pleistocene fauna. Tool and fire use had their origin in Africa, and here the ecosystems had the opportunity to shift to the superior hunting ability of the human race. But as migration brought one or another of world ecosystems into contact with these superior hunters, they had little chance to adapt. At the time of human arrival in Europe, 300,000 years ago, many Pleistocene species were still alive, including the forest elephant, hippopotamus, and rhinoceros, and the giant Maltese vulture. All of these became extinct between 250,000 and 100,000 years ago. In North America, the Pleistocene fauna fell even more rapidly—in a period between 11,000 and 8,000 years ago. Carbon datings of the remains of early North American man show that the bones of the last Pleistocene animals are associated with the hearths of the first invaders. In the La Brea tar-pits of California lie an extensive array of bones dating from 18,000 to 4,500 years ago. Of 54 species of mammals identified in the tar-pits, 24 are now extinct, and of 113 birds 22 are now extinct. They include the dire wolf, the short–faced coyote, the sabre-toothed cat, the giant lion, the super-camel, the American mastodon, two species of mammoths, several species of ground sloths, the asphalt stork, the La Brea turkey, and five giant birds of prey including Meriam's teratorn with a weight estimated at fifty pounds and a wingspread of twelve feet.

Until about 1,000 years ago, the Pleistocene fauna of New Zealand was almost intact. Giant moa birds occupied most of the niches belonging in continental areas to grazing animals. As well as twenty species of these birds, there were two varieties of flightless geese, a great swan and a great eagle. About 950 A.D. the first Maoris arrived in New Zealand. When Captain Cook arrived in the late eighteenth century, he was just in time to see the last moas.

Even though the Pleistocene era ended at about the same time all over the planet, the fall of the large Pleistocene faunas in each continent corresponds closely with the arrival of the first humans. There is, otherwise, no adequate reason to explain the extinction of these hundreds of species. They were well adapted to a world fluctuating between glaciation and warm interglacial periods. The sudden manner in which they disappeared is not at all characteristic of evolutionary change, and it varied with the technological superiority of the invaders. In Europe it required 100,000 years, in North America only 3,000; on the limited terrain of New Zealand it took less than 900 years, and on small islands in the Indian Ocean, where Pleistocene species included the dodo bird, it occurred within a century.

It is difficult to tell today the reasons for this over-kill. The species that suffered were often large and present in great numbers. They were, therefore, relatively easy to hunt and provided good yields of meat. Pressures of overpopulation may have been a factor, but this is mere speculation. Over-kills, however, are certainly not restricted to the history of the Pleistocene. One has only to consider the violence directed against the last North American remnants of the Pleistocene during the past 300 years—the bison, the passenger pigeon and the Carolina parakeet—to wonder if the human race has always displayed a deep-rooted irrationality towards its environment.

Agriculture and Domestication

The origins of agriculture are thought to have been an intensive form of food-gathering—the search of vegetation for edible roots, tubers, berries, shoots, and foliage. It is thought to have begun in the Fertile Crescent area of the Middle East, about 11,000 to 9,000 years ago, for it was in these areas that the ancestors of modern cereals such as wheat and barley are found. The basis of agriculture was the discovery that these seeds could be stored and sowed in a desired location. Even this first agriculture had a distinct effect upon the plant communities that were being cultivated. The seeds that were planted were gathered from plants that stood upright, and held tightly to their seeds. Although these plants could not reproduce well, for they were unable to disperse their seeds, they were easy to harvest. And they gave rise to plants that were also easy to harvest but unable to reproduce naturally. Thus today, the wheat plant is totally dependent upon the human race for the dispersal of its seeds. (A parallel situation has occurred in North America where the corn plant is completely dependent upon humanity.)

The domestication of animal species may have begun in the same period (or a bit earlier) as the domestication of plants. The dog was probably first, and likely in a hunting role. It may have provided early hunters with a keen nose and an inherent hunting ability, while the humans provided the dog with shelter and warmth. Domestication of herbivores such as the sheep may have begun with the use of young captive animals as decoys. A continuing process of breeding these animals in captivity and augmenting them with tamer individuals from wild herds selected for the more subservient animals. Again, like wild grasses, wild sheep and goats are poor at producing the milk and wool valued by humans. Domestication and selective breeding led to new varieties, and in time these may have crowded the natural species. Furthermore we find the development of a gene pool of qualities that were not necessarily suited to survival within a functioning ecosystem but were of maximum advantage to the human animal. Domestication was leading to a simplification of the food chains. Humans were consciously competing against such other top predators as the wolf for herbivores such as the sheep, towards which they held protective feelings. A conscious antagonism to the wolf may date from this point. Non-domesticated herbivores would be pressured away from the grazing lands.

As agriculture and domestication spread, the lands under human domination increased, reaching as far away as North Africa and China. Apart from regions like northern Europe where deep soils and gentle rainfall are permissive, hot seasons and strong winds combined with agricultural forces and overgrazing led to substantial soil erosion. Recent studies in Great Britain indicate that before the arrival of agriculture most of the island (including what is today heath and moorland) was covered with deep brown soils maintained by thick temperate forests. Primitive agriculture and overgrazing turned some soils dry and chalky, while others became leached and acid, and all over the country they prevented the forces of succession from renewing the soils. In the original Fertile Crescent, erosion has eaten away soils as much as fourteen feet, and the lower valleys of the Tigris and Euphrates have become almost completely infertile.

Domestication of animals may have increased human contact with parasites such as the trychinosis roundworm that lives in pigs. It may have increased the incidence of diseases such as tuberculosis that infect milk. But the first agricultural technology increased tenfold the amount of food available to humans from the biosphere. No longer dependent upon the produce of a single season of the year, the human population could store food and animal fodder during the winter months. In the light of greater food production, the effects of increased disease were hardly noticeable. Between the end of the Mesolithic era and the middle of the Neolithic 4,000 years later, the human population increased from five million to eighty-six million.

Agricultural Villages

Following the first agricultural practices came the first villages, probably communal centres out of which individuals could work the fields and to which produce was returned. The emergence of small agricultural villages was an important step for cultural evolution. Food production had increased and was more efficient, consequently the total work-hours necessary for food-hunting and gathering decreased. Thus, within Neolithic societies, many individuals began to play new cultural roles. People could remain at home cooking or preserving food and raising children. Older people were no longer an undue burden on the food supply and could be supported by the community. They represented a store of wisdom and experience which at this point the community could begin to afford and to encourage.

(Even in Neolithic times, the average human life-span was thirty years!) Many of the old people thus remained in the villages and passed their knowledge on to the children. It is still possible to see this social arrangement in cultures about the world. Other occupational transitions—and, again, ones that stimulated further cultural and technological change—came into play, with the advent of arts such as weaving, pottery, and basket-making.

Metallurgy and the First Cities

Gold was apparently the first metal to be worked, but it was rare and used mostly for ornamentation. Copper, on the other hand, was easy to obtain as well as easy to work, and it appears to have been the first metal put to widespread use. Copper utensils date from as early as 3000 B.C. in Mesopotamia. Tin was discovered later and combined with copper to give the alloy bronze. It was not until 1000 B.C., however, that furnaces hot enough to smelt iron were finally constructed and the Iron Age began.

All of these processes were limited in their scope and probably had only small effects upon the environment. Indeed, even as late as the eighteenth century many European foundries produced as little as two tons of iron a day. In these early times far less was produced, and the use of metals rarely extended beyond the vessels, jewellery, and weapons of the privileged classes. Agriculture for a long time depended on implements made of bone, wood, and clay. Some resource deposits were depleted however, and Egypt imported copper ore from as far away as Scandinavia.

Between 3000 and 2500 B.C. the first city states emerged in Mesopotamia as agricultural distribution centres based about the temples. Some of these cities contained up to 50,000 individuals, small in comparison with modern megalopoli, but large in terms of the Neolithic world. Although the rich and the commercial areas of the cities were laid out in orderly fashion, and even had buried conduits for sewage disposal, the majority of people had cramped and dingy living-quarters. Sanitary problems arose from the disposal of human and food wastes, and particularly through contamination of drinking water. Food and excrement would often accumulate in the streets, and the incidence of disease must have increased over that in the earlier farming communities. Furthermore, unsanitary conditions stimulated the invasion of *inquilines* like the mouse, the rat, and the fly-organisms that have adapted to life on the refuse of human communities. On the black rat, for example, live fleas that can carry the bubonic plague bacterium. These fleas, in moving from rats to humans, from humans to humans, and from humans to rats to humans, act as vectors in the transmission of plague. With increased human density, disease and parasites were on the rise, applying new forces of control upon human population size.

Transportation

With the development of wheel and sail, people could travel and carry materials over great distances. Plants, animals, and disease organisms were also transported, often without anyone's awareness. Living organisms were introduced into ecosystems that often had no mechanisms for their control. What changes accompanied the spread of trade routes across Asia, Europe, northern Africa, and the Middle East, it is today almost impossible to tell. However, from the experience of explorers and colonists, sent about the world from the fourteenth century onwards, we have ample evidence of the sort of disruptions that can occur and that probably took place in these earlier times. For example, in the grain taken to the Americas by European settlers were the seeds of many other plants. These, like the chicory, spread rapidly across the continent, displacing native plants. All over the hilly country east of the Rockies from Idaho to California, the inferior cheat grass (both a fire hazard and unpalatable to herbivores) has replaced the native grasses integral in maintaining the rich soils. Indeed most of our commonly recognised 'weeds' appeared in this manner in the New World. The introduction of the prickly pear into Australia was another disastrous move; vast areas have been overcome by this straggling and thorny plant. Plant diseases have also been transferred; introduced from Europe, both the chesnut blight and the Dutch elm disease have ravaged North American forests and more will come.

Birds and mammals have also been moved. One major disaster was the introduction of the rabbit into Australia in 1859. In Australian ecosystems the niches occupied in Europe by small mammals such as the rabbit are empty or partially filled with the kangaroo and other grazers. These organisms appear to be essentially self-regulating, and there are few top predators necessary to keep them under control. Within six years of being released into this ecosystem, the rabbit population had exploded from 24 individuals to 30,000! By the 1900s this population had climbed to the millions and in 1953, despite efforts to control it, had reached one billion. Hordes of rabbits competed at the water holes with sheep (another introduction), and together they overgrazed the grasslands, accelerating soil erosion. Today the rabbits have been at least partially controlled with the potent myxomatosis virus. In New Zealand the introduction of European birds has displaced many of the native species. In North America the starling—another introduction—is out of control.

Transportation systems have also spread disease organisms. Ships from continental Europe carried the bubonic plague via fleas on the black rat to England. In every port at which a ship docked, there was the possibility of disease transfer, either by people or by other vectors. Some medical historians believe that syphilis was brought to Europe by the sailors of Columbus. And when the Europeans came in contact with the peoples of America and the South Seas, they were surprised at the severity with which disease organisms could be transmitted. Smallpox, diphtheria, and even diseases such as measles, which among Europeans was a relatively common occurrence and of only mild effect, became potent killers among the highly susceptible peoples of Polynesia. Except for syphilis, there are few records of Europeans catching and spreading major pathogens from these sparsely populated native peoples. Here then is evidence that in urbanized Europe and Asia disease incidence had become an important control mechanism of population.

Population Trends to 1850

Between 1 A.D. and 1850 the population of Europe fluctuated considerably, but over all it increased. Three factors appear to have played important roles in the control of population size: disease, warfare, and famine. Their effects were sporadic and would hit a large part of the population in a sudden manner. Following this they might subside, leaving that part of the world in a relatively quiet state for a while. All three of these factors seem to have been intensified by high human population density.

In mediaeval Europe, disease was a particularly potent factor. The Black Death swept across Europe on two major occasions and appeared sporadically all across the Eurasian continent well into the eighteenth century. In the thirteenth century, it reduced the population of England by a third, and although its effect in the sixteenth century was less, it still claimed about 10 per cent of the population. Besides plague, there were a host of other killers: pneumonia, diphtheria, whooping cough, tetanus, smallpox, typhoid, cholera, childbirth fever, malaria, and influenza. Together these diseases gave rise to very high levels of infant mortality and to very low average lifespans.

War was another factor that appears to have been important. As far back as historical records date, war has been a common incident among humans. However, in a stable ecosystem, as far as can be seen, war of the intensity that occurs between men has rarely, if ever, been observed between members of any other species, particularly mammalian ones. It is possible that warfare may have arisen with urbanization and increased population densities, even perhaps through the disappearance of tribal living, but we cannot be sure of any of these explanations. War has however, until recently, had a dramatic effect upon the size of human population. Among those that affected the population of Europe in the Middle Ages were the Hundred Years War (1337 to 1453) and the Thirty Years War (1618 to 1648). In the latter an estimated one-third of the people in Germany and Bohemia lost their lives.

The availability of food was also becoming an important control factor. Records and inscriptions from ancient civilizations in China, India, Egypt, Greece and Mexico are dotted with references to famine. During the first fifteen hundred years A.D., famines affected most parts of Europe, and in Great Britain alone there were more than 200 recorded famines. One of the most recent took place when the potato blight hit Ireland in 1845.

In the mid-seventeenth century things began to change. In Europe the Thirty Years War ended with the Peace of Westphalia, bringing more stable times. New agricultural practices bringing increased yields came into common use—practices like strip farming and crop rotation. Also about this time, the brown rat (which inhabits sewers) began to displace the black rat (which inhabits houses); thus contact between rats and humans and consequently the incidence of bubonic plague was lessened. Fires like the great one in 1666 that destroyed much of London helped to raze the rat-infested wooden structures that had survived since mediaeval times. Newer structures were made of stone and decidedly more rat-proof. From 1650 to 1750 the populations of Russia and Europe increased by almost 25 per cent.

In Asia, up to the seventeenth century, things were much as they were in Europe. Famine, warfare, and plague were all taking heavy tolls and population remained relatively stable. For example, it has been estimated that in India over the two thousand years from 300 B.C. to 1600 A.D. the population remained at about 120 million.

In China the Ming dynasty fell in 1650 and was replaced by the Manchus who brought with them newer and more productive agricultural policies. By 1750, China's population had increased some 65 per cent. India's population remained relatively constant; the fall of the Mogul Empire in the mid-sixteenth century and invasion by the British and French led to a period of instability and famine—in the great famine of 1770 an estimated three million people died, including a third of the population of Bengal.

By the eighteenth century, invasion and settlement of the New World by Europeans had begun. Pressures on food sources in the Old World began to lessen and the shipment of agricultural produce from the Americas began to supplement home-grown crops. In Europe urban density was rising, and by 1750 the Industrial Revolution was sweeping England. By 1850 it had reached much of the rest of Europe. Living conditions began to rise slowly among some classes and very quickly among others. The first widespread public-health programs had their beginnings in these years.

Up to 1850, human population was still held closely in check by potent mechanisms—disease and famine. A long history of technological innovation had removed many of the factors at work on biological populations, including climate and predation, and had increased the food base for the human population. The widespread environmental changes wrought cumulatively upon the surroundings by this history of change had little meaning for people; and life in a biological system was too remote for anyone to remember. The changes in species composition of the biosphere, the eroded soils, the disappearing forests, and the extinct mammals and birds were an unrecognized part of the everyday world. For the majority of people alive in those times, particularly in the urbanized centres of the seventeenth and eighteenth centuries, life was a misery. But the tools and the methods that had brought about the changes to that point were an exciting suggestion of what was possible in the future for a greater portion of the populace. And one of these technologies, modern medicine, was to have far-reaching effects.

CHAPTER 2—2

Epidemic : The Human Animal : Phase Three

The mechanics of the human population, like those of any other organism in the biosphere, involve a complex interaction of many control factors. These range from parasitism to disease, climatic pressures, and predation through to competition, territorial behaviour, and the availability of food. The relative importance of each of these factors as a controlling influence varies with the life-style of the individual organism in conjunction with the state of affairs in the local ecosystem at a particular moment. With human population, the impact of technology on these controlling factors has had important consequences in determining their relative importance during various stages of human history.

We have seen how both tool-making and the agricultural revolution relaxed the limitations of food resources, and how shelter, fire and clothing afforded protection from extremes of climate and from predators. The dramatic burst of human population growth in the last hundred years is in large part the consequence of medical technology, reducing what had become one of the most stringent of all the population-control mechanisms—disease.

Figure 22—1 shows the growth of human population from 12,000 years ago to the present. The curve is roughly a logarithmic one, characteristic of any population that is growing rapidly. It is interesting to note that from the beginning of this graph to about the birth of Christ, human population grew at a very small rate. In fact, if one were to extend the bottom edge of this growth curve backwards some million years to the start of human history, then the line would extend almost forty feet from the edge of this page. In other words, for this great length of time the human population was remarkably constant.

By about 1800, the scientific world had acquired the tools, skills, and outlooks that were necessary for a successful attack upon disease. Microbes had been linked with the decomposition and souring of foodstuffs, and Appert had developed the first methods for the sanitary bottling and canning of foods. By the 1870s disease had been linked with bacteria by Pasteur and programs of quarantine and immunization were underway. Disinfection was becoming a widespread practice in hospitals and operating rooms, and by the turn of the century sanitation was extending to the chlorination of water supplies and the pasteurization of milk. With the advance of organic chemistry as a science, research into drugs and drug action began. Penicillin was discovered in 1929 and was followed by a host of antibiotics like streptomycin, aureomycin, tetracyclin, and the sulfa drugs. By the year 1970, thousands of chemicals useful in therapy against fungal and bacterial infections, and mental, physical and nervous disorders, had been determined. The list of diseases against which people could be immunized included smallpox, typhus, rabies, yellow fever, diphtheria, whooping cough, poliomyelitis, tuberculosis, tetanus, typhoid, paratyphoid, and bubonic plague. In tropical areas, pesticides like DDT were effective in the control of mosquitoes and other insects—vectors in the transmission of diseases like malaria, yellow fever, sleeping sickness, and infant diarrhoea.

The dynamics of a population may be expressed in a number of ways, most commonly in terms of the crude birth-rate and the crude death-rate. These are simply the numbers of births and deaths per

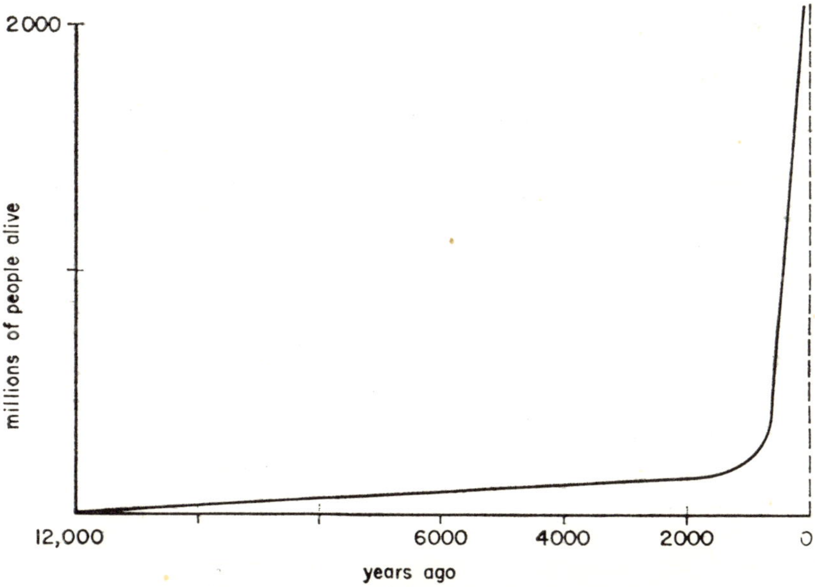

Figure 22—1 Human Population Growth

year per thousand people in a population. They are termed 'crude' because they are essentially untreated data and fail to give any breakdown of births and deaths according to age. The difference between the crude birth- and death-rates is the number of people added to, or taken from, the population annually, per thousand people. Dividing this number by ten will give the percentage annual rate of change of the population.

Medical technology has affected these rates in two ways. It has first of all reduced infant mortality and the number of stillbirths. Secondly, it has substantially altered the death-rate as it affects the older portion of a population. In the seventeenth and eighteenth centuries, up to one-half of all the children born died in their infancy (between years one to four). Today this figure has dropped significantly, for example, to 0.03 per cent in Sweden and 4 per cent in the United Arab Republic (1960-1962 data). This dramatic drop in infant mortality has been generally due to innoculation programs against smallpox, diphtheria, and whooping cough, and reduction of deaths due to childbirth, as well as to the widespread introduction and acceptance of sanitary measures and the treatment of infant diseases.

Death-rates in the other segments of society, from teen-age to middle-age, have dropped precipitously as well. With the advent of medical treatment the lifespan of an average Westerner has increased from forty years in 1460 to seventy years in 1950. Again this has largely been the consequence of successful disease control and, in parts of the world, of better nutrition. Between 1900 and 1950 in the United States the crude death-rate dropped from 17.2 to 9.6 per thousand. In 1900 the major causes of death were tuberculosis, pneumonia, and infant diarrhoea. In 1950, however, they were cancer, heart disease, cerebral haemorrhage, and accidents—predominantly the diseases of old age.

In total then, while the birth-rate has been rising slightly or remaining constant, the death-rate has been decreasing. More people are staying alive for longer periods of time, and more people reach reproductive age. Changes in population size and structure like these are termed *demographic transitions* (demography is the study of human populations). They have varied about the world with social conditions and with the level of industrialization.

In western Europe for example, a slow-down of population growth followed the Industrial Revolution. Marriage occurred at a later average age and the number of children per family decreased. Large families in urban areas had an entirely different economic meaning from the large families of an agricultural society. In the cities, it was better to have fewer children. And with urbanization in twentieth-century Europe came the first family-planning programs. Thus, in Europe the impact of medicine in changing population growth-rates was less dramatic than elsewhere in the world. Nevertheless, between 1850 and 1950 the population of Europe more than doubled in spite of massive emigration.

At the same time, the U.S.A. and Canada were largely agricultural nations, in which large families predominated, and were encouraging immigration to the New World. Attitudes towards family planning were repressive. Consequently between 1850 and 1950, the population of North America increased sixfold.

Very sharp demographic transitions occurred in Asia in this period. Between 1750 and 1850 population size increased 50 per cent. But in India and Pakistan alone in the period 1850 to 1950, it trebled, without the high migration rates that gave such a boost to North American growth. (Annual birth-rates over this period remained relatively constant at forty-six to forty-nine per thousand.)

In Ceylon, an intensive program of DDT spraying was introduced in the year 1946 to control the malarial mosquito. Within two years of the inception of this program the death-rate had dropped from twenty-two per thousand to 13.5 per thousand. By 1969 the death-rate had dropped again to eight per thousand, and the greatest decline in this rate was observed among the young. Whereas in Sweden between 1900 and 1969 the rate of growth of population decreased from 1.0 to 0.8 per cent, in Ceylon it increased from 1.0 to 2.4 per cent.

Similar results followed the introduction of medical health programs into many countries. Between 1940 and 1950 the death-rate dropped some 46 per cent in Puerto Rico, 43 per cent in Formosa, and 23 per cent in Jamaica. On the average, in the underdeveloped nations it dropped 24 per cent. Control of diseases such as typhoid, paratyphoid, typhus, malaria, yellow fever, and cholera has been a large factor in these changes. Between 1850 and 1900 the population of Africa increased 20-40 per cent and by 1950 it had doubled. Between 1850 and 1950 the population of Asia almost doubled, and the population of South America increased fivefold.

In the million years that humans have walked the earth, about 110 billion people have been alive. On the average, in any year of human existence there have been about 110,000 individuals. Today there are 30,000 times that number, almost 3 per cent of all the people that have ever lived!

Projections of Population Growth

Before considering how projections of future population size are estimated, let us consider how populations of living organisms grow. Let us take a completely sterile test-tube containing all of the nutrients necessary for the health of a microbe. Then let us take just two of a particular organism and place them in the tube and seal it. The life-cycle of these organisms happens to be ten minutes. In other words, given adequate food supply each cell will adsorb food from the medium and grow enough in ten minutes to reproduce (by mitotic fission) into two daughter cells. The population will have grown from two to four organisms. Ten minutes later this will re-occur and there will be eight microbes in the tube. Thus the doubling time for the population is ten minutes. The population will continue to grow in this manner until it either runs out of food or is poisoned by its own wastes.

This sort of growth is typical of a population from which many of the control factors have been removed. The doubling time may be ten days, or ten years. It is dependent upon the reproductive ability of the population in question. For the human population, the doubling time is now thirty-five years. Mathematically, this mode of growth is termed a logarithmic or exponential one. (Indeed for the human population, growth has become what is termed compound logarithmic; since the doubling time has been decreasing as the rate of growth has been increasing. The resultant growth curve is steeper than a purely logarithmic one where the doubling remains constant.) The mathematics of exponential functions provide methods of making population projections, using an accumulated background of statistics and trends. These projections can be made under a range of different assumptions that attempt to determine the effects of family planning, preventive medicine, and other factors upon population growth. To make such projections, finer details of the population are needed, and a careful evaluation of the influences at work within the population must be made. Let us consider some of these, beginning with the age-structure of a population, the number of individuals in each age-group.

Future population trends will be affected by the percentage of fertile individuals in a society and the percentage of young people who, although not yet fertile, will be within the time of the estimate. These people will be an important influence on the projection. In a society that consists largely of individuals beyond the reproductive age-brackets, future trends of growth may be much less than in one where nearly 50 per cent of the society is thirty years or younger. In other words, the birth- and death-rates of a society can change

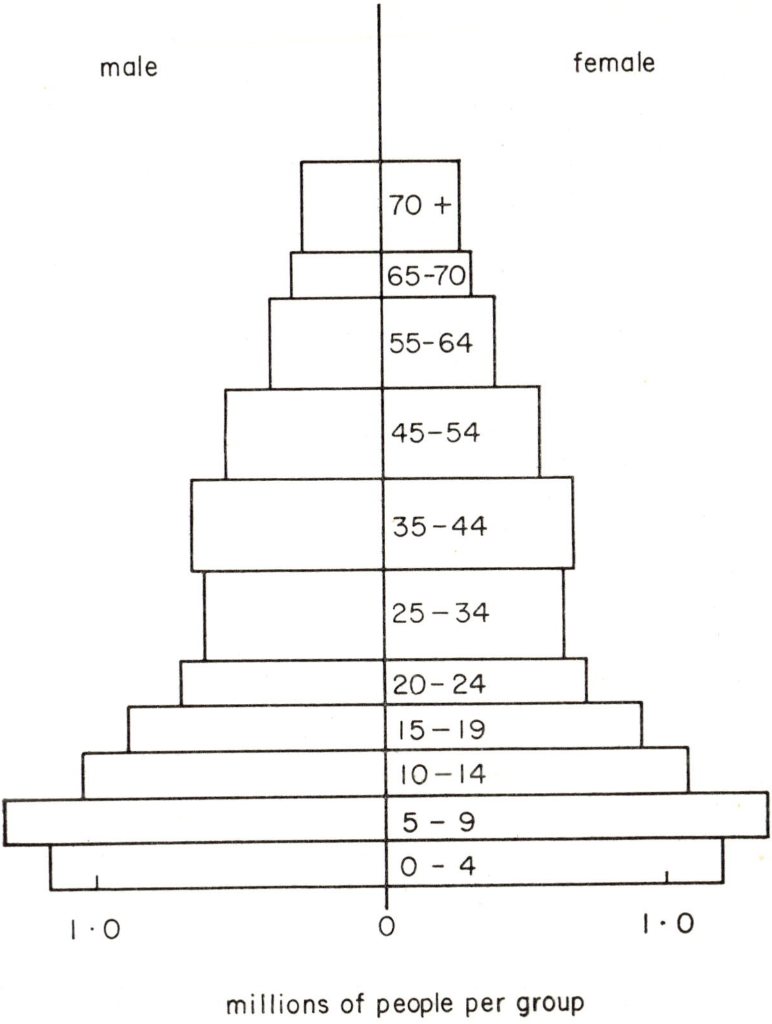

Figure 22—2 Age-Structure of the Canadian Population (1966 Census)

over time, and these changes must be accounted for in a population projection.

Because social and economic conditions have a large influence upon family size, the age-structure of a population often reflects the level of economic development attained by a nation. In Mauritius and India, for example, high birth-rates and declining death-rates place the largest percentage of society in the one to twenty year-old age-bracket (particularly since reduced infant mortality plays a large part in the lowered death-rates). This distribution is characteristic of a rapidly growing population and is illustrated in Figure 22—2. (Canada is a major exception among the developed nations, with an age-structure typical of a developing nation.) The more rapidly growing the population, the broader the base of the pyramid, and the narrower the peak.

In a slowly growing population like that of Great Britain, there is a very different distribution of age. A much larger part of society is in the upper age-brackets, and there is a more even distribution among the lower ones. In other words, the number of young people is not increasing rapidly, so that, migration aside, the future population size will remain relatively close to what it is now. The age-structure of the Japanese people is an interesting one. In age-brackets above ten years, it is typical of a quickly growing population; below that age, there is a sharp decrease in numbers. The concerted efforts of the

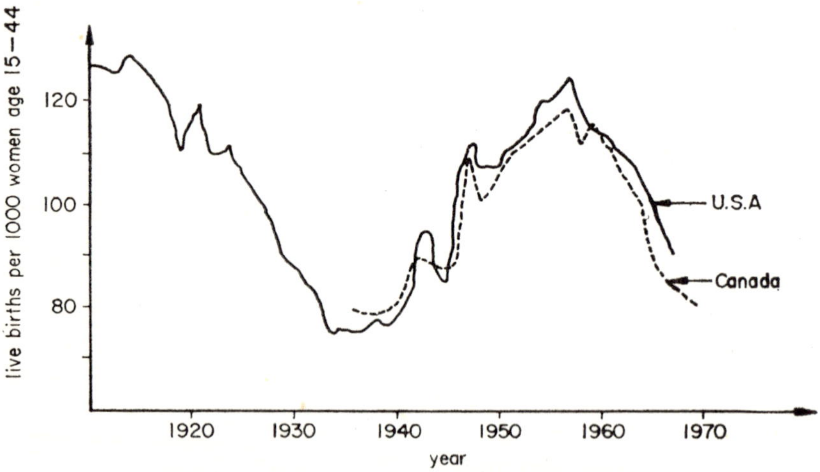

Figure 22—3 Fertility Curves, Canada, U.S.A.

Japanese to slow population growth are showing in fewer children, and once the large group of potentially reproductive individuals today have passed maturity, population growth should begin to level out. Early indications suggest that this has happened.

Age-specific birth- and death-rates change dramatically with social conditions and attitudes, and these may have profound effects upon population predictions. In the United States and Canada, for example, fertility rates (live births per 1,000 women in the fifteen to forty-four age-bracket) rose and fell with peace, war, and economic conditions (Figure 22—3). During and following the Great Depression, fertility reached a record low. Following the Second World War and during the subsequent economic boom of the fifties it climbed dramatically, dropping again during the sixties as the number of women of child-bearing age (twenty to twenty-nine) fell (these were the women born in the low of the depression). This does not of course signal a drop in the population growth—from 1965 onwards the women born during the post-war baby boom have been reaching their reproductive years.

Immigration rates are another important factor that can have far-reaching effects upon population projections. They are dependent to a large degree upon government policy. As we have already noted, they played an important role in the dramatic rise of the North American population between 1850 and 1950, and in Canada between 1851 and 1966. Here, immigration rates varied from 2.2 to 3.6 per cent greater than the native growth-rate.

Government policy aimed at depressing or stimulating growth-rates of population can also have various effects. In France for example, following the Revolution of 1798, migration from the rural areas to the cities began. The traditional practice of a farmer subdividing his land among his sons had reached practical limits; further subdivision would only yield farms that were uneconomical. During the nineteenth century rural migrants began to accumulate in the cities where jobs were difficult to find. Because of this scarcity of land and jobs, the birth-rate of France fell substantially, and incentives by the government to increase it again have failed.

In countries such as India, Pakistan, the Philippines, Columbia, and Chile, governments have taken steps to control population growth. Family-planning services have been introduced on a widespread and voluntary basis in order to bring the birth-rate down. It has been argued, however, that if efforts of the French government to increase the birth-rate did the opposite, then efforts to decrease population may, as a result of unpredicted social and economic factors, lead to an increase. At this point it is difficult to predict what the results of these programmes will be, but they do appear to be having some success. In certain countries such as Thailand and Jamaica, birth-rates have begun to fall quite rapidly; however, population growth is still more than 2 per cent per year.

Because of the great complexity of interaction of the factors that can affect population, rather than making one projection alone, it is a sounder practice to make a broad range of projections. By making each of these under a number of specific conditions (for example, constant fertility and decreasing mortality), one can attempt to cover the whole range of future possibilities. A number of world projections made in this manner are shown in Figure 22—4. They cover a range

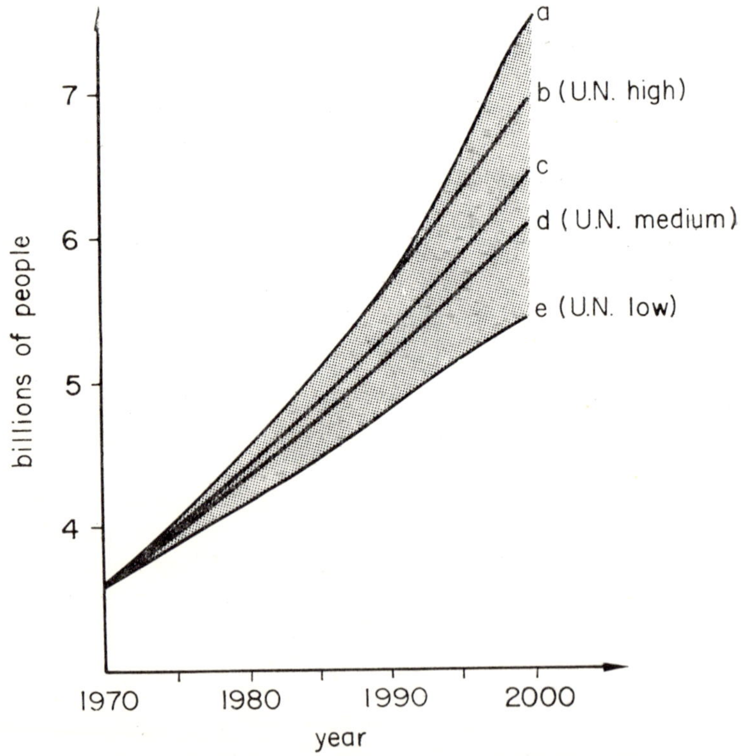

Figure 22—4 World Population Projections

(for the year 2000) of 5·4 billion to greater than 7·5 billion. The highest prediction assumes constant rates of fertility and a declining rate of mortality. Should mortality and fertility remain constant, population would increase at 2 per cent per annum, giving curve C. Curves, B, D, and E are three predictions made by the United Nations, termed high, medium, and low respectively, each assuming a certain decrease in fertility due to family planning.

All of these are only projections of present trends, and they may or may not turn out to be the same as future growth. But if past experience is any indication, then they are likely short of the mark. Time and again population projections have been made, only to be boosted a few years later when better evidence became available, or when actual population growth showed that the projections were underestimates. The combined population of Pakistan and India was projected for the year 2000 in both 1951 and 1966. The 1966 estimate was 1·2 billion, up by more than a third of the 1951 estimate of 650 million.

The U.N. high estimate predicts that world population will double its 1970 level just following the year 2000. Should the U.N. medium estimate hold then it would just reach the six billion mark, and should this growth-rate continue into the twenty-first century, then human population would cross the ten billion mark by the year 2050. This growth can be further divided into the rates of growth in the developed and the developing world:

Table 22—1 U.N. Medium Estimate, Growth-Rates of the Developed and Developing World

Region	Billions of people	1970	2050
Developed		1·08	2·04
Developing		2·51	8·32
Total world		3·59	10·36

By the year 2050, should these trends hold true, then the developing world will have increased from two-thirds to four-fifths of the total population. The world already faces a crisis in attempting to provide underdeveloped nations with a higher quality of life; this situation will only grow tighter with continued population growth. Should we fail even to reach the degree of control necessary to follow the U.N. medium projection, then world population may climb as high as eighteen billion in the year 2050, although there are many ecological reasons that suggest a size this large could never be reached.

Zero Population Growth

A population has reached a state of zero growth when the birth rate plus the immigration rate equals the death rate plus the emigration rate. In Italy for example, the population growth is 2 per cent and the net migration rate is −2 per cent per annum. In other words, Italy has zero population growth.

In another way of looking at it, zero population growth is achieved when each couple on the average has only two children. It is necessary to take the average because a certain percentage of these children will never reach child-producing age, and a certain percentage of the population will neither marry nor produce children. However, should attempts be made now to achieve zero population growth, such a rate will not be reached for at least another generation. As the age-structures of growing populations predict, there is a greater proportion of persons in the young age-bracket than in any other group. On the average it will be another twenty-five years before these people have matured and produced children, and therefore until population will actually begin to level off. This lag period before population-control methods become effective is termed the *braking time*. On a world scale, should we be successful in the next five years at introducing world-wide and effective family planning, it would be another thirty years before we would achieve zero population growth. Unless there is world-wide catastrophe in the next thirty years, world population will probably reach at least the seven billion mark by the year 2000. Should we be unsuccessful in this decade at beginning zero population growth, then we are faced with an even larger increase in world population before a levelling-off is possible. This is the crisis that we face: as long as we continue to delay taking action in the fullest and most meaningful way, we will be faced with a worsening situation, which Nature may find soluble only through efficient means such as plague, mass warfare, and ecological catastrophe.

Population Growth and Urbanization

Density of human population will also change with population growth.

In particular, rural-to-urban migration patterns will change. Just as it is possible to project our present knowledge of population growth into the future, it is also possible to estimate the changes in population density over the next fifty to a hundred years. In the last twenty years the rural-to-urban migration rates have been very high; should these trends continue we shall be in for some major problems. By 1984 about one-half of the world's population will be living in cities of over one million inhabitants, and by 2024 the total world population will be living in cities of that size, with the largest city having almost one and a half billion people!

How these cities will manage to cope with this growth is another question. Calcutta with 7·5 million people now is estimated to reach sixty-six million in the year 2000, yet today several hundred thousand people live homeless in the streets. Providing even educational and sanitary facilities for these people will be a major problem, and one that is already gripping South American cities. Thousands of squatters move annually from the rural areas of South America to set up shanty towns (where living conditions are pitiful and health hazards extremely high) about cities like Lima and Sao Paulo. Similar shanty towns grip the world from Shanghai and Hong Kong to Rome, Mexico City and Jerusalem.

In the developed nations, urban renewal presents a problem of great magnitude and cost. In cities across North America, Europe, and Japan, crime-rates are rising as more people are crammed into urban areas, for a large majority of whom conditions do not improve. Conditions are aggravated by the design of cities, so that the majority of suburban and well-to-do individuals have little knowledge of how the underprivileged live and, for the most part, care even less.

Canada is a prime example of the modern trends in urbanization. Of seventeen major cities, all except six (Halifax, Saint John, St. John's, Victoria, Windsor, and Winnipeg) will have to accommodate twice as many inhabitants by the year 2000. Toronto and Montreal will double in size, and five other Canadian cities will have over a million people. Even today, uncontrolled urban sprawl about cities like Vancouver and Toronto is destroying good agricultural land. Already there is a shortage of near-by recreational facilities, and many cities are falling victim to the automobile and the expressway mentality. Between 1970 and the year 2000 the percentage of the Canadian population living in urban centres is projected to rise from 74 per cent to 94 per cent. Can governments keep up with the financing of necessary housing and facilities, at the same time maintaining a high quality of life?

Justifiably in many people's minds, those trapped in the North American urban world have begun to break out in violence against the immobility of poverty, as the hordes of Bombay and Calcutta occasionally have broken into rioting over food. These problems will not decrease; they can only continue to worsen with the increased population growth and with the predicted increases in rural-to-urban migration rates.

Conclusion

Except for cancer, by the year 1970, medical technology had achieved a resounding success in curing the physical ills of the body. By the year 1970, it had done much to alleviate human misery and suffering. It had reduced childhood afflictions that had previously doomed many individuals to an adulthood of incapacity. On the whole, human health had improved dramatically. But human health cannot really be judged apart from environmental health, and medical technology must be measured up against the biological processes that it has altered. In tampering with the disease mechanism of population control—particularly in the potentially explosive world of the mid-eighteenth and early nineteenth centuries—medicine had merely exchanged one form of bondage for another. As human population exploded, environmental quality began to decrease and particularly in the underdeveloped world, famine struck with an impact, the magnitude of which had never been felt before.

Today, human population is climbing upwards against all the fundamental limits of the biosphere and the Earth. We are facing a crisis in agricultural, mineral, and energy resources, among many others. In the use of all these resources human technology is having detrimental effects upon the biosphere. These effects, and the basic resource limits against which human population is now crashing, will be considered in the following chapters.

CHAPTER 2—3

Agriculture : The Green Dream

These beans have results which are not harvested by me. Do they not grow for woodchucks partly? . . . Shall I not rejoice also at the abundance of weeds whose seeds are the granary of the birds? It matters little comparatively whether the fields fill the farmer's barns. The true husbandman will cease from anxiety, and finish his labour every day, relinquishing all claim to the produce of his fields, and sacrificing in his mind not only his first but his last fruits also.

Henry Thoreau, Walden, 1847

We have already explored the first processes of agriculture. Today they are much the same. Ecological structures are simplified and food chains are shortened so that a majority of the productivity is directed towards the human race. Animal and plant species that either have no immediate nutritive value to humans, or are obvious competitors, are stamped down. The process of simplification has even extended to the soils where the concerted action of plant and decay organisms produces both richness and long-term stability. Our early attitudes towards competitor organisms and our general ecological ignorance today are perhaps reflected in our common usage of the terms 'weed', 'pest', and 'vermin'.

Today these processes of simplification have become stronger and more forceful, and the ecological imbalances that they induce have become more evident. Modern science is oriented towards the elimination of competitors and the expansion of monocultures. And in an attempt to solve the instabilities that result from the simplification of ecosystems, we are turning to technology: pesticides, fertilizers, and irrigation. All of these, however, tend to aggravate the situation rather than improve it, for they lead to still greater simplification and still greater imbalances. In the long run a vicious circle is springing back on us. We are reducing rather than increasing productivity, and our dreams of changing a green world to an even greener one are likely to become a nightmare of infertile deserts and eroded countryside.

The Soil Ecosystem

Soils are a result of the interaction of climatic forces (winds and rain), biological systems—particularly plant and decomposer communities—and the geological nature of a region. The products vary with all three of these factors. The action of wind and rain is one of breaking up rock material into finer particles through the forces of erosion. High winds and heavy rainfall will increase erosion. Vegetative cover will tend to slow these forces down by providing protective cover. Plant roots, however, despite holding together the substrate material, continue to break it down. We observed earlier how the process of succession, given adequate nutrients and climatic conditions, can transform an area of barren ground into one of climax vegetation. In the process, each plant community in the continuum lays down a cover of dead matter which is further converted into soils and held in place by the plant root systems. As the vegetation increases in size, so does the protection from wind and rain, and the system becomes self-maintaining, new nutrients being laid down while others are removed.

Soils under a plant community are never static. Ground water is continually percolating through them. As it passes over the substrate rock it dissolves minerals, which it carries into the soils. Conversely, as ground waters seep out of the soils, they carry minerals with them, and rainfall brings with it dissolved carbonates and nitrates from the atmosphere. However, because the water movement is slowed down by soil ecosystems, and because many of the plant nutrients are tied up in a complex form with the decaying matter, soils tend to maintain their moisture and nutrient value. A study on the annual loss and gain of calcium in one temperate forest ecosystem showed that the annual losses of calcium through erosion about equalled the annual input from weathering and rainfall. Stable soil ecosystems, particularly those under climax communities, appear to conserve nutrients, either by trapping them within the soil litter, or adsorbing them to clay and humus particles. The retention of scarce nutrients like phosphate appears to be stronger than the retention of commonly available ones like potassium and sodium. However, in all cases there is some retention, and removal of vegetative cover and disruption of soils during agriculture and forestry have shown just how important this action really is.

By breaking the soil into chunks, the plant roots expose soil surfaces to the atmosphere so that they remain aerated. Burrowing organisms like the mole and the earthworm contribute greatly to this process, and this action is important, for the decomposer organisms that inhabit these regions are dependent upon atmospheric nitrogen and oxygen. In layers farther down, where oxygen levels are lower, one finds the anaerobes such as the sulfate-reducing bacteria. Near the surface are the nitrifying and denitrifying bacteria. Potential habitats for soil organisms are provided in the spaces between surface litter, in caverns between soil aggregates, in root channels and fissures, and in pore spaces between individual particles of soil. These micro-environments vary in temperature and moisture and aeration. In this very complex and stratified environment live the organisms producing and maintaining stable soils.

Decomposer organisms form an interesting series of relationships among themselves in their own right constituting a balanced ecosystem based upon the detritus or dead-life matter. Table 23—1 shows the relative numbers of some of these organisms in forest and meadow soils.

Table 23—1 Some Populations of Soil Organisms*

Organism	Average number of individuals/ 1,000 cubic centimetres
Protozoa	1,000,000,000
Rotifers and tardigrades	500
Nematodes (roundworms)	30,000
Springtails	1,000
Mites	2,000
Other small arthropods (spiders, insects, millipedes, and centipedes)	100
Earthworms	2

*Data from Friedrich Schaller, *Soil Animals* (Ann Arbor: University of Michigan Press, 1968).

Figure 23—1 shows a partial map of some of the food chains that exist in many soil ecosystems. Population-control mechanisms again occur in many forms. Some depend solely upon food, and bacterial

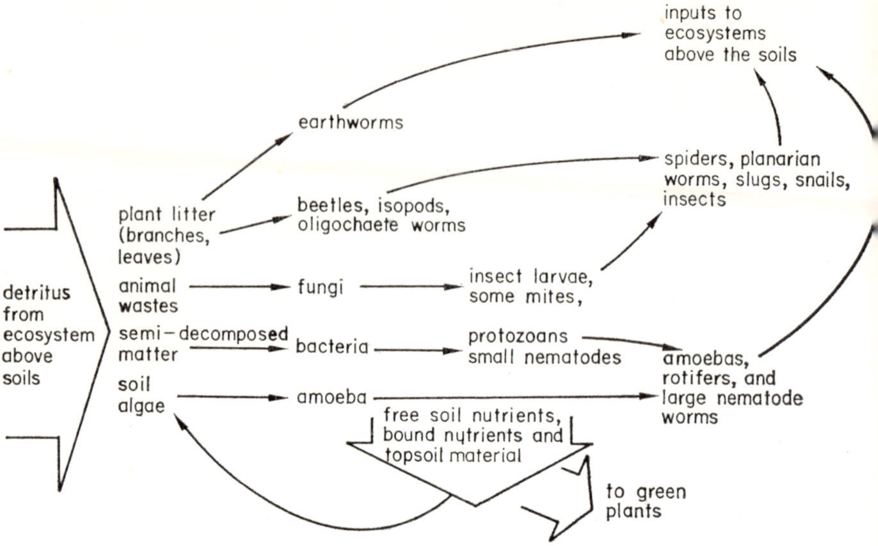

Figure 23—1 A Simplified Soil Ecosystem

populations rise and collapse in response to its availability. Protozoans act as predators upon bacteria and certain worms feed upon protozoans. Small insects, worms, and mites are preyed upon by spiders and larger insects.

Decomposer organisms act together in a concerted manner upon the material entering soil ecosystems from above. Let us follow, for example, the decomposition of a fallen leaf. Holes are first eaten in the leaf by springtails and bark-lice, then by insect larvae. The combined action of wood-lice, earwigs, silverfish, millipedes, and dipteran (fly) larvae continues this process until only the vein-skeleton of the leaf remains. With all the actions of all these organisms, the leaf is converted to tiny fragments with a much-increased total area of exposed leaf. Hence microbial action climaxes at this point, with yeasts, fungi, and bacteria taking part in the processes of fermentation and decomposition. Earthworms passing through the soil ingest the small leaf particles, along with clay and mineral particles in the soil. The material excreted from their intestinal tracts contains aggregates of clay and humus—the beginnings of topsoil formation. (This process is repeated many times through the guts of the earthworms and their relatives.) Finally the digging and burrowing of small animals loosens and reshuffles the mixture of humus and mineral particles to form an organic mull—the body of a rich topsoil. The formation of topsoil is a process that takes millions of diverse organisms working for hundreds of years.

The soil ecosystem that we have just discussed is an ideal one, most typical of communities in temperate, semi-temperate, and semi-tropical areas. These are areas in which rainfall is neither so heavy that vegetative cover is of little avail in preventing erosion, nor is so light that plant growth is inadequate to exert any control over the soils. In the former cases one finds rain-forests, in the latter case, deserts. Tropical rain-forests have adapted to high rainfall, with its strongly erosive action, thus maintaining some of the highest levels of productivity in the world. Topsoils here are thin or non-existent. High temperatures and humidity provide conditions under which soil bacteria and fungi thrive. Acting immediately and quickly on litter, they convert it to soluble nutrients that are washed into the soils and are soon re-absorbed by the plant roots. Soil invertebrates are scarce in rain-forest soils. Towards the edges of the Amazon jungle, where rainfall is less forceful and temperatures lower than in the heart of the jungle, soils become richer and more stabilized with the formation of substantial humus layers. Similar conditions exist in the humid cloud-forests on the peaks of tropical mountains and on the west coast of North America. Deserts, on the other hand, are unproductive, since the lack of rainfall limits growth to highly adapted species such as the cactus. Plants are widely dispersed and as a result can exert little environmental control. The nutrient value of the soils is, because of the low rainfall, largely unobtainable by the plants, and is washed out by sudden sporadic cloudbursts which create deep eroded gulleys because of the few roots holding back the soil.

The organisms of soil ecosystems, just as organisms of any ecosystem in the biosphere, have been selected through years of evolution for their roles in the cycles of decomposition and soil maintenance. They are well adapted to the ecosystems above them and follow successional patterns with them. Plant, animal, and decomposer communities are inextricably linked together through millions of years of co-development, and this partnership is a key factor in biological stability. In other words, the soil ecosystems of the fertile and stable climax communities cannot be expected to function optimally without those communities; for the most part, the stability of any soil ecosystem reflects the ecological diversity and productivity of the community above it.

Problems of Agricultural Technology

1. Loss of Nutrients and Soil Material

When the protective systems of plant roots are removed and the soils are ploughed, irrigation by rainfall and by ground waters is much freer. The water-retention ability of the soils is reduced and the nutrient material carried out annually by the waters is very high. Because of the geometric spacing of crop communities and the removal of 'weed' species, the crop does not replace the dense network of root systems provided by the original community.

Table 23—2 shows some of the annual losses under a number of crop covers:

Table 23—2 Loss of Nutrients from Fields under Various Crops* (lbs. per acre annually)

Crop	Nitrogen	Phosphorus	Potassium	Calcium	Sulfur
Continuous bluegrass	0·6	0·2	2·7	1·1	—
Fallow (ploughed 4″)	118	38	1246	459	47
Continuous corn	66	18	605	221	17
Continuous wheat	32	9	264	106	9
Continuous wheat-clover Rotation	26	6	214	86	6

*Data adapted from A. F. Gustafson, *Soils and Soil Management* (McGraw-Hill, 1941).

Continuous bluegrass best approximates the original biological conditions of the protective cover. The ploughed land left unplanted (fallow) exemplifies the worst possible sort of thing that can happen to the soil ecosystem and here losses are greatest. Yet this is a practice farmers have adhered to for centuries. Of the crop systems, although in every case nutrient loss is much higher than with the intact system, continuous rotation of two different plants maximized the retention of nutrients.

Not only nutrients are lost at a greater rate under agriculture, but also particles of humus. Without the protective network of plant roots, these are easily removed by fast-flowing surface waters (especially after heavy rainfall and in irrigated areas), and by wind erosion. Table 23—3 shows the losses of both soil and water under a number of different covers.

Table 23—3 Soil and Water Losses under Various Vegetative Covers*

Crop	% water	lbs. per acre of soil
Meadow (fertilized)	0·7	35
Fallow (ploughed 4″)	17·0	22,231
Corn (fertilized)	8·3	7,009
Rotation (corn, oats, barley and clover)	3·1	509
Idle (weeds and some clover)	2·5	884

*Data adapted from Gustafson (see Table 23—2).

During this study, the May to October rainfall in all of these plots was twenty-four inches. Again, the best cover for maintaining soil particles and the moisture content of the soils was the meadow, with conditions nearest to a natural ecosystem. Leaving the land bare and ploughed accelerated the loss of soil particles almost a thousandfold and decreased the water content of the soils by a third. In a similar study, the times required to lose seven inches of good topsoil were estimated for a number of vegetative covers. Land left fallow suffered erosion of this magnitude in only twenty-five years. For a cover of continuous bluegrass 3,000 years were required. A rotating system of corn wheat and clover would reduce the life-time of the seven-inch topsoil to about 350 years.

With the loss of water from the soils and removal of extensive plant root systems, the properties of soil noticeably change. Greater exposure to sunlight accelerates water loss through evaporation, and causes caking. The particles of soil thus clump together and become hard and dry, losing their aerated quality and becoming powdery and susceptible to erosion. Much of the plant material is removed annually as crop and is not replaced in organic form. For all these reasons, the micro-environments of the soil change, and with them the types and numbers of soil organisms. In Europe, for example, in land under cultivation with sugar beets, the number of earthworms was less than a third of the number in permanent meadowland (a reduction of 172 to 47 per square metre). Other organisms also diminish in number. In one study comparing cultivated meadowland, one-fifth of the number of nematode worms, one-sixth of the number of mites, one-sixth of the number of springtails, and one-tenth of the number of enchytraeid worms (small relatives of the earthworm) were found. Bacterial and fungal populations also change drastically in numbers and in species composition in soils under cultivation.

Ploughing up a field, then, and subjecting it to a monoculture will heavily dampen many of the mechanisms that renew the nutritive value of the soils and that preserve their unique physio-chemical qualities. The practice of crop rotation overcomes some of these problems by partially changing the micro-environments each season, so that a greater number of fertility-restoring organisms have a chance to reassert their effectiveness. For example, including leguminous plants like alfalfa in the rotation introduces colonies of nitrogen fixers in the soils for a year of the cycle. Leaving the land as pastureland for a year provides a new layer of dead organic matter and permits earthworm, fungal, and bacterial populations to process this material into humus.

Agriculture, in a short-term effort to create a greater human-oriented productivity, is destroying the long-term value of the soils as stable, fertile, and self-maintaining communities. In global and historical perspective this is very frightening. Areas of China and Brazil have suffered heavily from poor farming practices introduced to grassland areas. The Gobi Desert of central Asia, areas of Utah, south-west Africa, western Texas, and Mexico have suffered similar damage through wind and water erosion resulting from poor land management, overgrazing, and the cultivation in many cases of areas that were just not suited to agriculture in the first place. Ancient civilizations also tell their story in this manner. In the Yangtze and Indus valleys, millions of people exploit the infertile and muddy soils, remains of hundreds of years of erosion. Erosion was stimulated a number of centuries ago in China when the emperors decided to increase land under cultivation and cleared the forested uplands. Subsequently the rains roared down the river valleys, washing everything in their path and choking riverways with the eroded silt from the uplands. As the plains became flatter with erosion, flooding became a common occurrence in these shallow valleys. In the late 1950s the U.S.S.R. set out on a broad programme to cultivate the 'Black Earth' steppe lands about the Caspian Sea. These lands were only suited at the most for a little grazing. Their attempts were subsequently met with wind erosion and disaster.

Perhaps the most dramatic example is the exhausted Mediterranean basin. Today we consider the climate of hot dry summers and cool wet winters, and the accompanying hot bare rock, sparse soils, and small shrubs and trees, as typically Mediterranean. Yet before the advent of ancient civilizations, the region was one of thick virgin forests that stretched from Lebanon (the famous cedars of Lebanon) to Spain. Through time these forests had built up stable and extensive soils, and conserved the waters necessary for survival through the hot dry summers. Leaf cover cut back on wind speed and provided shade, thus lowering evaporation. In the wake of many empires the forests gradually shrank, to be replaced with agricultural land, their lumber providing wood for ships and even the piles on which Venice is built. The agricultural land shrank with erosion from water during the winters, and from wind during the summers. The Babylonian, Egyptian, Phoenician, Greek, Roman, Moorish, Spanish, and Italian empires all took their toll. The last region to go was that of Illyria on the Dalmatian coast, at the time of Venice's height as a naval power. The introduction of the goat to these areas proved fatal, for its browsing action, combined with the fury of the north wind during the winter months, removed any but the toughest and thorniest of new vegetation. Malaria became epidemic in these areas about 1,800 years ago, perhaps because the heavy erosion accompanying Roman farming in these regions created silty low-lying wet areas that were ideal for mosquitoes. Consequently people moved to the higher areas building hill-towns and thus accelerating the erosion further.

In eighteenth-century Europe, awareness of some of these problems was beginning to spread. New and saner agricultural practices such as crop rotation (alternating with pastureland), strip-farming, and contour ploughing were introduced. However this was possible only at the expense of the recently colonized New World. Here, agricultural land appeared infinite. When an area was exhausted, people could always move farther west. And this was the way that the pioneers went, cutting the surface cover and ploughing up the land, planting it and harvesting it, taking little or no heed of wind or water erosion. The ultimate results can still be recalled by many today—the dust bowl of the American mid-west in the 1930s.

In tropical regions many of the soils are of a peculiar nature termed laterite. These soils are very difficult to manage, and no technology is known today to make them as manageable as temperate soils. When the vegetative cover is cleared and the soils ploughed and exposed to the air, their constituents undergo complex chemical changes, becoming rock-hard. In Brazil, within five years of clearing cover, soils were virtually concrete-hard pavements. Presumably, similar events led to the decline of the ancient civilization at Angkor Wat in Cambodia. Areas of disrupted laterite soils can be found in West Africa, India, south-east Asia, Australia, and South America.

2. Irrigation

Irrigation is an ancient agricultural practice, in which water courses are diverted over areas of arid and semi-arid land in order to make them arable. A number of problems arise. Lack of plant cover and extensive root systems accelerate erosion of both nutrients and soil particles. In coastal regions salinity gradients may be upset. These are the gradual transition zones from salty to salt-free land; they are maintained by a washing-out action of both ground-water and rivers. Under irrigation, ground-water tables often rise and the salt water is carried upwards, turning the topsoil salty and rendering the land unsuitable for agriculture. In the Babylonian valley, irrigation dams have been impounded with silt, and salinization of the lower reaches has occurred. Salinization has destroyed agricultural land in Pakistan and, ironically, also in the Nile delta as a result of the new Aswan Dam, built to provide electricity and irrigation to Upper Egypt, and in the valley of the lower Colorado River. And in the United States today more than 2,000 irrigation dams are impounded with silt.

3. Fertilizers

The loss of nutrients from agricultural systems has been countered by the addition of fertilizers. These are mixtures of inorganic materials that are generally derived from rock, or else chemically synthesized. Sometimes they include biological material such as manure or bone-meal. High-phospate rocks, potash, and sulfate rocks are combined with industrially fixed nitrogen in the form of ammonium salts or nitrates. Good fertilizers, then, provide sources of sulfur, nitrogen, phosporus, sodium, calcium, and magnesium, as well as trace quantities of essential elements.

The soils of an ecosystem, however, are adapted to an annual inflow of dissolved mineral that is relatively low in comparison with the nutrient content of the soils (note Table 23—1). The nutrient content of the soils tends to be a mixture of both free, inorganic nutrients and organically combined nutrients in the form of decomposed matter like humus. Nutrient in this form is probably much more valuable to green plants than nutrient in the form of fertilizers. In the soils an equilibrium exists between the organically combined nutrients and the small amounts of free and soluble nutrients. A slow release of free nutrients occurs over time and at a rate concurrent with steady absorption by plant roots. Applying inorganic fertilizer to a field where the organic content is low does not bring about the same results. Fertilizers may be washed out easily, before they are needed by green plants; and often the increased addition of fertilizers, beyond a certain point, will do little to increase crop yields.

Furthermore, the presence of large quantities of inorganic fertilizers appears to have detrimental effects upon some soil organisms, particularly the nitrogen-fixing bacteria. There is even speculation that excessive use of these fertilizers could push some of these bacterial species to local extinction. Fertilizers appear to inhibit soil-building processes, with a decrease in the nitrogen content of the soils and in the formation of humus. (The ability of green plants to absorb nitrogenous compounds seems to be directly related to the amount of humus present.) We know far too little of the complex roles of soil organisms and of the intricate biogeochemical pathways involved in the maintenance of soil fertility to be certain of what the long-term effects of heavy and widespread fertilizer usage will be, but one estimate gives a decrease of 50 per cent in the original nitrogen content of mid-western soils in the U.S.A. through the use of fertilizers. If this is at all representative of the degree to which inorganic fertilizers are affecting long-term fertility, we may be in for some rude awakenings regarding their true effectiveness.

4. Biocides

GENERAL CONSIDERATIONS. In a balanced biological system, the presence of predatory and disease organisms prevents the populations of plant-eating organisms, except under rare circumstances, from reaching epidemic proportions. In agricultural systems, however, this is not the case, and for two major reasons.

First, the agricultural system provides an ideal banquet for the pest organism. Travel from one plant to the next is made easier by the neat and close arrangement of the plants and by the absence of other plant types.

Secondly, in an agricultural system there is less diversity of habitat —tree trunks, undergrowth, wet places, etc.—than in a functioning ecosystem. Thus there is a reduction in the number of plant-eating organisms and their predators. Some plant-eating insects, for example, will have predators present so that their grazing action upon the crop never crosses the level of economic damage. Other insects may be in the alternate situation, having few predators and growing madly out of control. Thus, an agricultural system is an ecosystem where one or more of the control mechanisms are missing. This can hardly be over-emphasized, for in the attempt to control the pest population, one must make certain that these efforts do not destroy the ecological balances that are already intact.

On a crop of tomato plants, for example, there may be two types of pests: a beetle and the tomato hornworm. The first is a major pest, while the latter is kept well under control by its predator a wasp. Indiscriminate application of a pesticide may destroy the beetle insect and wasp, but not the tomato hornworm that is resistant to the spray. Uncontrolled, the hornworm will now become a major pest. Pesticides, then, are bad strategy in the first place and should not be used until a careful understanding of their effects both inside and outside of an ecosystem is achieved. Many of the problems that we face from pesticides today are a consequence of such indiscriminate use.

Other problems develop upon the use of pesticides, one of these being resurgence. The application of pesticide may wipe out most of the pest insects, but the few survivors, finding little or no competition for food, may literally explode back to their original density through rapid and unhindered reproduction.

Resistance is another problem, and in some cases it has developed so rapidly that chemists have had difficulty countering with the development of new pesticides. In Canada alone, well over a hundred major pests have developed resistance to certain pesticides. Resistance is the ability to detoxify a pesticide either by combining it with another chemical to make it harmless, or by breaking it down. The ability to do this is selected for in the following manner.

We have a population of beetles feeding upon cabbage plants. A pesticide spray known to be toxic to this beetle is applied repeatedly. The population, except for a few individuals, is wiped out. These few individuals could withstand the spray, for they could detoxify it. It is these few survivors that breed to produce the new population—many or all of whom, like their parents, will be resistant to the spray. This is evolution in action. Developing resistance in this manner requires a massive population wipe-out with the survival of only a few organisms, and survival is only possible because the breeding-rates and reproductive cycles of the pest organisms are very rapid.

There are good alternatives to the indiscriminate use of pesticides. These are based upon an understanding of the complex factors surrounding a semi-ecosystem such as a cornfield, and are termed *integrated control*. They will be discussed more fully in Chapter 3—2.

SPECIFIC PROBLEMS OF BIOCIDES. Each type of biocide that is employed has specific problems associated with its use in the environment. They relate to the innate chemical nature of the biocidal molecule which can generally be classified as long-life or short-life biocides.

LONG-LIFE BIOCIDES. The biocides of this type received their name because of their persistence in the environment, and it was for this prolonged effectiveness that they were highly valued. The most famous of these types is DDT, a chlorinated, aromatic, hydrocarbon derivative (1,1,1-trichloro-2,2-bis-(p-chlorophenyl) ethane). A host of chemically similar biocides accompany DDT, including aldrin, dieldrin, heptachlor, toxaphene, lindane, isodrin, chlordane, and endrin. As we discuss DDT in the following paragraphs, it is important to remember that much of the discussion is also relevant to these pesticides having similar properties. Each of these pesticides combines the four important physio-chemical properties exhibited by DDT.

(i) DDT is very stable in the environment. It is metabolized only very slowly by living organisms to degradation compounds called DDD and DDE. Both of these are similar to DDT; the former, however, is less toxic, while the latter is more toxic, than DDT. In all it is estimated that DDT and its degradation products have a half-life of at least fifteen years. In other words, should ten grams of DDT be released in the environment today, in fifteen years five of those grams would still be around in the form of DDT, DDD, or DDE. Consequently, there has been an accumulation of DDT in the environment from the time of its first use about thirty years ago.

(ii) DDT is highly insoluble in water, and very soluble in hydrocarbon materials. Thus, since life is comprised of hydrocarbon matter, there is a tendency for DDT in the physical environment to become absorbed into the body of living organisms and to concentrate in areas such as the fatty tissues that are richest in hydrocarbon. A minnow, for example, will absorb DDT from the water through its gills and its skin, and from the food that it eats. More DDT will be absorbed and retained than is excreted. A gull feeding on that minnow will absorb all of the DDT in that minnow and excrete very little of it. Because the gull feeds on many minnows, it absorbs all of the DDT from these organisms, concentrating this pesticide within its own tissues. Thus, per unit of body weight, the gull has accumulated more DDT than the minnow. This process occurs at every level of the food web, so that by the time DDT reaches the top predators it has been concentrated as much as a million times over the amount in the environment. Table 23—4 illustrates this process, known as *biological concentration.*

(iii) DDT is very mobile. It becomes adsorbed to soil particles and is thus washed into streams, lakes, and the oceans. When sprayed from a plane, less than 50 per cent of the DDT actually reaches the ground. The rest is dispersed through the atmosphere, some of it being carried

Table 23—4 Biological Concentration of DDT[1] in a Long Island Estuary†

Ambient environment	Producers	1° Consumers	2° Consumers	3° Consumers
Water (0·00005)	Cladophora (0·03)	Shrimp (0·16)	Blowfish (0·17)	Merganser (22·8)
Bottom muds	Plankton (0·04)	Silversides (0·23)	Billfish (2·07)	Ospreys
-marsh (13lbs/ac)	Marsh plants	Fluke (1·28)		-eggs (13·8)
-bay (0·3lbs/ac)	-shoots (0·33)	Eel (0·28)		Green heron (3·6)
	-roots (2·80)	Mud snail (0·25)		Terns (3·25-5·2)
		Clam (0·42)		Cormorant (26·4)
		Minnow (0·94)		Gulls (3·5-18·5)
		Cricket (0·28)		

[1] DDT+DDD+DDE is given in parts per million.

† Adapted from data in G. M. Woodwell, C. F. Wurster, Jr., and P. A. Isaacson, *Science,* 156, 824 (1967).

all over the globe by air currents. When DDT does reach the ground it does not stay there, for it codistils with water—in other words, as water evaporates, DDT is carried with it. DDT is also carried thousands of miles in the flesh of migrating birds and fish. DDT has been found in the ice of Greenland glaciers and in penguins and seals in the Antarctic.

(iv) DDT has widespread biological action. It accumulates most in the fatty tissues and it is here that its effects are primarily felt. These tissues include, as well as the adipose tissues, the brain, the nerves, the liver—site of fat metabolism—and the gonads. As a nerve poison, DDT appears to act at the nerve-junctions. Seagulls containing large amounts of DDT were starved, simulating conditions during winter or long migration when the birds are forced to tap their fat reserves. Large amounts of DDT were brought in contact with their nervous systems and the gulls died from erratic and berserk behaviour similar to that observed in poisoned insects.

High levels of DDT have been blamed for the failure of bird and fish eggs to hatch, or fish fry to live past the first few days following hatching. Minute amounts of DDT have been shown to affect the ability of fish to avoid electric shocks or to find their way through a maze in search of food.

Furthermore, DDT has been shown to affect the activity of one of the liver enzymes responsible for metabolizing the sex hormones. In other words, the activity of this enzyme is a crucial factor in the sex-hormone balance, and DDT destroys this balance. In birds this balance is crucial to the amount of calcium laid down on the eggshell as calcium carbonate. Nest surveys and museum studies have established a sharp decrease in eggshell thickness for many predatory birds such as the osprey and the eagle. Because of the thin shells, the eggs lose their moisture and warmth quickly and are easily crushed by the mother. Lack of reproductive success is imperilling many bird species from gannets and pelicans through hawks and owls. DDT also appears to affect this same liver enzyme in laboratory mice. What the insidious effects of this are in the human population is presently impossible to tell. Recent work has also shown that DDT is carcinogenic.

In summary, DDT and its relatives are spread over the biosphere from the initial site of their application by a large number of means. Spraying in South Africa affects Canadians as spraying in Alberta affects Malaysians. There is a continual flow of DDT from the reservoirs of the physical environment—lakes and ocean muds, and soils and waters—to the food chains where it concentrates, to be felt most heavily by the top predators. In the fatty tissues of humans, DDT concentrations range from two ppm in Great Britain to ten to twelve in Canada, nineteen in Israel, and a high of thirty-one in India.

A wide spectrum of other biocides have become long-life contaminants of the biosphere. Mercury is an important one, for organo-mercury chemicals are used to treat seed-grains against fungi and bacteria. In certain areas of the world, widespread mercury contamination has resulted, and in western Canada for example, populations of predatory birds such as the prairie falcon are endangered. Other long-life biocides include dibutyl tin and compounds of arsenic, but very little is known of their effects. For example, arsenic is insanely used for the harvesting of potatoes. Because the potato foliage gets in the way of the harvester, half a day or so before harvest the leaves of the plants are sprayed heavily with arsenic. Several hours later the plants wilt and the machines can move forward unobstructed, turning the vegetation and the arsenic into the soils as they go. Nothing is known of the effects of the long-term build-up of arsenic used in this manner in the soils of areas such as Ontario's Holland Marsh. Arsenic in the form of cacodylic acid is used extensively by the U.S. government in South-east Asia to kill rice crops.

One other long-term biocide, which is extremely potent and again is employed in the defoliation warfare in Vietnam, is picloram. Areas sprayed with picloram have remained defoliated for two years. Then, when donkeys fed upon the sparse new growth, their manure was collected and stored for a further year. When used as fertilizer this manure still had potent defoliating powers.

SHORT-LIFE BIOCIDES These materials are degraded quickly by the biosphere, either by atmospheric or geological conditions or by the degradative metabolisms of various organisms. Their end products seem to be molecules that the biosphere is able to comprehend. Because they degrade quickly, they have only a short time in which to work and their toxicity must be very much higher than that of the long-life biocides. Consequently, very small amounts of pesticides like azodrin and parathion can be fatal to mammals. Great care must be exercised in the usage of these materials, and there are many possible hazards to which spray operators are subject. Foods treated with these materials must be allowed to sit for a prescribed period of decontamination.

Other than the immediate acute toxicity of these chemicals, little is known of their long-term sub-lethal effects. For instance, what are the effects of exposure over a long period to repeated sub-lethal doses? Electroencephalographic studies of workers in pesticide industries suggest that brain and nervous functions may be impaired.

Herbicides like 2,4-D and 2,4,5-T have been linked with malformed births in Vietnam where these chemical have been a major part of the war effort. These chemicals have also been used broadly across North America. It appears, however, that the mutagen is not 2,4-D itself, but a common contaminant of 2,4-D, 2,4,5-T, and other biocidal agents like pentachlorophenols. The contaminant is termed *dioxin* and is one of the most potent teratogenic (causing genetic malformations) chemicals known.

The major forms of short-life biocides include the organophosphate (azodrin, malathion, phosdrin, parathion, and dichlorvos [Vaponal]), the carbamate (Sevin), and the isocyanate insecticides, and the chlorophenoxy acetate (2,4-D and 2,4,5-T) herbicides.

The Food Crisis: Meeting the Limits of the Earth

It is evident that agriculture has costs in both ecological stability and in the fertility of the soils. Already we may, through bad agricultural techniques, be destroying more arable land each year than we create. Only through careful agriculture, on a sustained basis (using a bare minimum of crop rotation), will we be able to maintain the long-term productivity of the soils.

Weighted against an urgent need to move towards less intensive agriculture are the demands of an expanding and underfed human population of 3·7 billion people. If the food production of today's world were divided equally among all people, on a protein and calorie basis, then it would fall just short of the United Nations' minimum standards. Most critical of all is the protein shortage. Even though agriculture today produces enough protein to supply every person with sixty-seven of the seventy minimum grams, this is considerably off the mark, for two-thirds of the world's protein comes from plants, the other third from animals. And plant protein is neither as digestible as animal protein, nor does it contain an adequate balance of the essential amino acids. The world food crisis, then, rests predominantly in a shortage of good-quality protein, and this shortage is heightened by a maldistribution of foodstuffs. On the average, a person in North

America consumes ten times the amount of animal protein that a resident of India does, and on the average ten times the U.N. minimum protein requirement.

In 1971 about one-seventh of the world's people, or half a million, are undernourished. An additional billion suffer from nutritional deficiencies, particularly in protein. The results can be seen in nearly every country, including the United States, where children exhibit the symptoms of kwashiorkor, or protein deficiency. Other common disorders arising from deficiencies are: beri-beri from lack of thiamine, rickets from lack of calciferol, blindness from lack of vitamin A, and pellagra from lack of niacin. No one appears to have calculated yet the world budget for these essential nutrients, but if we are barely scraping on calories and are insufficient in protein production, then it is unlikely that the situation is satisfactory here either.

With the annual growth of human population, agricultural advances are barely managing to maintain per-capita consumption levels, let alone improving them. In what directions can we turn in an effort to meet this crisis? Many solutions and panaceas have been offered, and they have been well discussed elsewhere (see Bibliography). A few of these are, however, worth mentioning here.

The most traditional methods appear the least promising. Of the estimated eight billion acres of arable land in the biosphere, three and a half billion are now under cultivation,* and for the most part these are the areas best suited for agriculture. The figure 'eight billion' acres is very deceiving, for included in it are areas that will only be developed with radical new technologies, a great deal of knowledge, or should never be developed at all. Included in that total are regions of laterite, semi-desert, and low-nutrient soils. Most of this land lies in tropical regions of Africa, South America and Australia. Of the total potential in South America, only 11 per cent is estimated to be developed. And if Brazil is any example of this potential, then the future is not very promising. Although the Amazon basin contains 59 per cent of the surface area of Brazil, the soils are either poor and leached by rainfall, or they are laterites. In the 160 million acres classified as potentially arable, only one-twentieth are above water during the rainy season. Of the southern grasslands only those of the two southern states have an agricultural potential equivalent to European soils. In the others the grasslands are sparse and nutrient-deficient, and towards the interior the subsoils are so hard they can be penetrated only with dynamite.

*United States Presidential Advisory Committee, 1967.

Time and again, great schemes of technology have attempted to bring tropical soils under cultivation. One project in northern Australia planned two million acres of rice. Within several years the project was cut by a third and still ran into financial and ecological difficulties. Another project, aimed at cultivating ground-nut (peanut) plantations in Tanzania, met with even more disaster. After six years' work and an investment of over $100 million the project was abandoned, leaving behind eroded and hard-packed soils. Extremes of heat and heavy sporadic rainfall, the special nature of tropical soils, and physical limitations such as distance, all bring about drastic failures in the use of conventional agriculture. Except for the fertile lowland and upland areas that are presently cultivated, and areas of volcanic soil which may offer some future potential, there are perhaps sound reasons why the tropics have remained agriculturally underdeveloped.

Tropical ecology has other aspects that are difficult to reconcile with traditional agriculture. Ecosystems are complex, highly productive, and extremely powerful biological forces. Vast numbers of insects, fungi, and parasites invade crop monocultures with devastating effects. Termites, weaver birds (the quelea), elephants, bats, snakes, predatory mammals, all take their toll. Extermination programs have had little effect and, if anything, have gone further in upsetting the ecological balance. In India, where human populations have wiped out most of the top predators, a vast army of termites, ants, elephants, parrots, antelopes, and rats compete with millions of human individuals for the produce of the fields. Irrigation systems, the draining of marshes, changes in the water-table of tropical regions all affect the whereabouts of disease and parasitic organisms. The silting-up of rivers in Africa due to irrigation and power dams creates conditions favourable to a type of snail. This snail bears a parasitic organism termed *schistosomiasis*, which burows in human flesh causing itching and discomfort all over the body. In Rhodesia alone an estimated nine-tenths of the population is thought to be infected.

The introduction of agriculture to areas that are presently uncultivated will require innovative and well-thought-out techniques if it is to succeed. These techniques must integrate with, and work alongside of, the ecological factors that presently exist. Otherwise these regions, as past experience has shown us, will deteriorate into a state that will not biologically, and certainly not agriculturally, regenerate. The erosion of valuable soils all over the biosphere is one of the greatest threats that we now face; we should have no desire to continue accelerating it.

Expanding world fisheries does not appear to be more than a partial answer to the food crisis. In a masterful account of the ultimate potential of the oceans, William Ricker, of the Resources and Man Committee, calculated a maximum sustainable yield of 150-160 million tons of fish products annually. Ricker feels, however, that a yield of 100 million tons is a more attainable estimate. The 1968 yield was 60 per cent of this, an indication that we are already reaching near to the limits. Furthermore, to approach this limit, fishing will have to place new emphasis on conserving the species that are being taken, for overly-intensive fishing is already causing declines in many catches of the most valued stocks all over the world.

The feasibility of harvesting other trophic levels of oceanic ecosystems appears widely varied. Phytoplankton themselves are dilute in concentration, and often unpalatable. Although their productivity is high, it is expressed through a rapid metabolism rather than a large bio-mass. Certain seaweeds and algae can be cultivated and are already used in nations such as Japan as a food source. Although seaweed-harvest might potentially supply as much as 2 million tons of food annually, this would supply but a fraction of the world's total demands.

By far the most promising directions in mariculture (marine-farming) are those aimed at raising large numbers of shellfish such as oysters on great barrages of racks lowered into enriched temperate and tropical waters. These are excellent sources of protein and tonnages of 10,000 tons per square mile have been achieved annually off the coast of Spain. Another feasible alternative is fish-farming, already used to some degree in South-east Asia to raise protein. Large yields (up to a hundredfold over coastal waters) can be attained by growing milkfish in ponds that have been fertilized with human sewage.

Better distribution is certainly an important direction that we can take. The rich anchovy catches from the coastal waters of Peru are ground into fish-meal and shipped to Europe, Japan, and America, to become fertilizer, catfood, poultry and livestock feed. Australia feeds the high-quality protein wastes from milk-processing to livestock; the amount of this protein each year would double the present per-capita levels in Japan. Both of these examples illustrate the sort of gross wastage that is taking place. With fishmeal, for example, a large percentage of the energy and protein value is destroyed in conversion to livestock. Along these lines, agricultural land could often be put to better use. Vast portions of tropical farmland are turned over to tea, coffee, cacoa, and bananas, most of which go to the developed world as luxury items.

New directions for the 'green revolution' may provide some relief for a short while. Its methods, however, are highly exploitive, and in the long run destroy the soils, creating even greater instabilities. Some of the attempts look promising, like the development of high-yield grains. However, great care must be exercised here because these new plants have been under surveillance for only a short period. They may prove to be highly susceptible to as-yet-undetermined insect pests and fungal diseases, or to climatic changes. Placing too great a dependence upon them could lead to even greater disaster. And finally, many of them are completely dependent upon fertilizers.

All of these directions will in the long run be futile, unless they are coupled with an all-out effort to level off population growth. To feed the yearly increment of world population with per-capita intake of protein enjoyed in the U.S. would require a 48 per cent increase in world fisheries. It is likely that the oceans would not suffice for much more than one year's increase. To maintain the present inadequate levels of food consumption between 1970 and 1975 will require at least a 15 per cent increase in world agriculture in Asia, Africa, and South America. To stay even on a per-capita basis from 1970 to 2000, production must double. Without population control, therefore, we are doomed to failure as our technological advances are swept away by an increasingly demanding population torn apart by outbreaks of famine.

The Harvest of Trees

Ecosystems and Fire

Before the arrival of the first European colonists, the vegetation of North America was a mosaic of communities. It stretched from the chestnut and oak forests of the Appalachians to the grasslands of Central America, to the Douglas fir and redwood forests of the West Coast. Each of these communities existed at the various stages of succession. Within a region of coniferous forest for example, patches of climax spruce were interspersed with patches of sub-climatic jack-pine and of new scrub growth that follows burning. Fire appears to have been the driving force of succession in most North American ecosystems. (Fire is not unique in North America as an ecological force, it plays important roles in many other world ecosystems including the savanna grasslands of Central Africa.)

Fire, it seems, is necessary to return the succession to its first stages. Because decomposition is slow in temperate regions, the communities often become overgrown and large quantities of organic matter accumulate in a semi-decomposed form (dead trees, branches, needles, and leaves). Fire releases the nutrient material again, carbon dioxide and some nitrogen to the atmosphere, and calcium, phosphorus, and other nutrients as ashes to the soil. North American ecosystems have adapted so well to the sweeping action of fire that it is an essential part of their existence.

Plants such as the jackpine depend upon fire to reproduce. It is not until the heat of the flames touches their tightly sealed cones that they open and release seeds. Across the grasslands and woodlands of North America are a wide variety of fire-resistant and fire-dependent plant strains, such as the fireweed, certain goldenrod species, milkweed, ticktrefoil, bush clover, lupine, Indian grass, and big bluestem. In particular, many varieties of leguminous plants with their nitrogen-fixing bacterial associates thrive following grassland fires. It has been said that the timber wolf and the deer are also dependent upon fire, the latter for the new sub-climatic growth upon which they feed. Indeed, ecosystems appear to have become dependent upon fire as an important mechanism of population control. The spread of spruce budworm, a common insect of spruce climax forests, appears to be limited by fire. The spread of mistletoe, a parasitic plant, seems to be similarly controlled.

Immediately following fire, many fire-dependent plant species begin anew the process of succession; following their settlement, invasion of other non-resistant species occurs. Burning of spruce climax forests is followed by a rapid growth of jackpine, birch, alder, and a number of herbaceous plants, all within a season. Thus the barren period following fire is minimized and erosion of topsoil and fire-deposited nutrients is diminished. Similar patterns following fire can be observed in prairie communities.

For a long time in North America, as elsewhere, the approach to forest management has been one of greenness is goodness. We have failed to conceive of the essential nature of fire as a force in the lives of North American forest ecosystems. Rather, we have termed it destructive, unnatural, and have attempted to stamp out virtually every fire that occurs. Certainly, since the arrival of the European settlers and particularly with the upswing of camping as a recreation in the last fifty years, there has been a greatly increased incidence of fire in North America. And because the areas of natural ecosystem have been drastically reduced, it is questionable whether total fire-regeneration is either a feasible or a desirable thing.

However, because of fire prevention, many North American forests have reached stages of succession that have rarely if ever occurred before; for example, the over-grown spruce climax forests of the Quetico area. There are no communities adapted to succeeding this climax, because succession has never before reached this point without ultimately facing the renewal of fire. Without it the bush becomes 'stagnant' with semi-decomposed vegetation and is visibly overgrown. Should fire catch hold in such a location, the intensity with which it would burn would create far more destruction than that in a normal climax or pre-climax fire. And because fire has been withheld, organisms such as the spruce budworm and the mistletoe have ravaged acres of forest and have become serious pests.

Overgrown climax forests are ripe for cutting, the lumbering concerns claim, and present this as partial reason for the necessity of logging in national and provincial parks. If we have reduced the incidence of fire, they say, then we shall have to introduce our own forces of renewal. And one of the most commonly used is clear-cutting, the removal of all the vegetation in a stand. These areas are then left to regenerate by succession. Clear-cutting, however, does not leave the community in the proper biological shape for resuccession in the way that forest ecosysems have adapted to it. It does not open the cones of the jackpines that are scattered about the spruce climax, nor does it stimulate the fire-dependent herbaceous growth of the first successional stage. Nor does it return the nutrient to the soils, for 50 per cent is left in the form of undecomposed leaves, bark, sawdust, branches, and crushed undergrowth. Succession then does not occur at as rapid a rate, nor does it follow the same patterns.

If forests are replanted following clear-cutting, they are not necessarily replanted with the best sort of tree, but with one type of tree in a very geometrical manner. Cut spruce forest will not be replaced with a mixture of jackpine seedlings, birch, and alder, but with one desired species. Consequently, the soil-building processes of succession will be reduced.

Forest Management

The forests of the world are a vital part of the productivity of the biosphere. They serve also as an important human resource—for lumber, wood-pulp, and recreation. The impact of human technology upon forest ecosystems has been as great as that upon any other ecosystem. Poor concepts of forest management have led to overgrown vegetation through the prevention of fire; depleted and eroded soils through forestry practices such as clear-cutting and lumber roads; the creation of monocultures in which pest organisms thrive; the displacement of native species of plants through the introduction of foreign ones; and the overcutting of forest land in order to meet the demands of a growing population and economy.

The consumption of wood and wood products in North America is rapidly increasing; concurrently there is a widespread depletion of forests. One estimate made in 1964 claims that, except under the most optimistic assumptions, the United States' demand by the year 2000 will push the forests of the east on the road to extinction and cause substantial depletion in the west through the cutting of growing and younger stands. Imports to the U.S. of Canadian softwood could reach as high as three billion cubic feet per year at the turn of the century except that Canadian excess will be barely two billion cubic feet. Today, massive lumbering across Canada, even in provincial parks, is being carried out to satisfy world demands. This lumbering is very much at an exploitive stage, and inadequate thought is being given to the future. Let us take British Columbia as an example.

In 1967, B.C. forests produced 77 per cent of Canada's softwood lumber, 100 per cent of the red-cedar shingles and shakes, 93 per cent of the softwood plywood, 46 per cent of the marketed chemical wood-pulp, and 14 per cent of the newsprint, paper, and paperboard. Of this material, 77 per cent of the lumber, 28 per cent of the plywood, 95 per cent of the shingles and shakes, 54 per cent of the pulp and 78 per cent of the paper were exported to foreign markets. And 40 per cent of Canada's foreign-exchange earnings from forestry products came from British Columbia alone.

The annual rate of growth of B.C. forests is estimated (by the B.C. Forest Inventory) at 2·3 billion cubic feet per year. This amount however does not represent the *net* annual growth, for there is an annual loss to fire, insects, and disease. A conservative estimate of this annual loss is 0·8 billion cubic feet. Thus the net growth-rate is about 1·5 billion cubic feet annually. In 1967, the amount of forest cut was 1·7 billion cubic feet (more than twice the cut of 1952), and the rate of cutting surpassed the rate of tree growth. Should lumbering continue to increase in rate as it has from 1952 to 1967, then in 1982 it will reach 3·4 billion cubic feet annually—more than twice the annual growth-rate. Although the total volume of B.C. forests is about 250 billion cubic feet of mature timber, with another 250 billion of immature growth, a rate of consumption of this level could deplete these forests, extensive as they may seem, within several centuries. This threat should be therefore not too difficult to correct today.

These estimates do not include losses due to clearing forests for agricultural purposes, and depletion from overgrazing. They do not include the fact that growth-rates will increase with the increasing percentage of younger forest, simply because younger trees grow faster than older ones. But they do point out that even in areas supposedly rich with tree-growth, our current demands are exceeding the ability of the natural world to replenish the resources that we extract.

Many alternatives to our present practices of forestry management have been suggested. One of these is to allow controlled burning to allow the natural cycles. This, however, would have to be restricted to true wilderness areas, and, furthermore, it destroys the timber potential of the climax stand. But it does solve at once many of the problems faced by forestry-management programmes. Controlled burning is proving successful in many forests of the National Parks system of the United States.

Since clear-cutting does not adequately substitute for fire as a successional force, greater research is necessary. Would limited on-site burning of the lumber scraps prove feasible? Another direction that holds great promise is that of slowing down succession, rather than attempting to prevent it, through selective cutting and careful, mixed replanting, of the type observed in Germany and Sweden. Perhaps a mixture of the two procedures will prove feasible, for example, selective cutting of the climax stand followed by controlled burning to induce succession. Certainly, replacing a climax community with a monoculture, as is often done, leads directly to many of the problems of agricultural systems, most typically the outbreak of pest organisms.

These practices must vary from area to area, dependent upon plans of land-usage. In parks, for example, succession through controlled burning should be encouraged, both because these are essentially biological preserves, where biological processes should be allowed to work unhampered, and because the noise, erosion, and lumber roads that accompany logging are against the basic value of the park as a biological and a recreational unit.

Human consumption of the renewable resources—food and forests—is proceeding at very high rates. These rates, in attempting to satisfy present demands, are mortgaging the human future. The situation with the non-renewable resources, however, is even more alarming.

CHAPTER 2—4

Technology and the Flow of Matter

An Overview of the Problem

Matter in the biosphere flows along closed and cyclic paths. It is through these cycles that life has shaped and is maintaining the stability of the environment on both a long-term and a short-term basis (Figure 24—1). At any point in time the biosphere is an equilibrium, so that flow along A is equivalent to output along B and C. Uptake along A is controlled by mechanisms within each ecosystem and by limiting factors. Waste conversion along C is an integral and balanced part of the system—not a burden—and is related to both population size and the availability of resources. Finally, the matter-flow system of the biosphere is characterized by a uniformity of molecular language—the molecular components of the biosphere are recognizable by it. In other words, biochemical methods

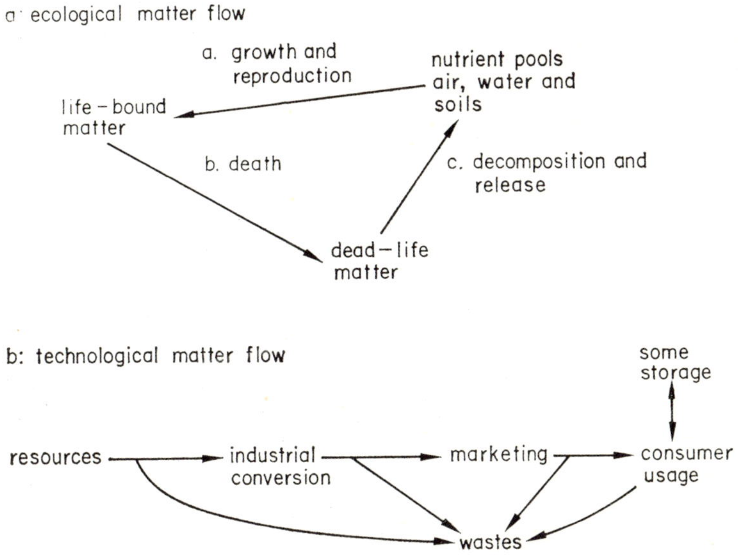

Figure 24—1 Ecological Matter Flow

exist for both the manufacture and the degradation of every biological molecule.

In contrast, the processes of technology are methods for converting matter into forms useful to the human race. Materials from within the biosphere (such as wood, plant oils, or atmospheric oxygen) and from without the biosphere (such as uranium, sulfur, or cobalt) may be transformed, usually along a pathway as illustrated in Figure 24—1. In other words, resources are converted to products, and along the way wastes are produced. Eventually the products themselves are transformed into wastes, whether slowly, as in the case of a concrete and steel building, or quickly, as an aerosol freshener is. The whole system, then, is one of producing wastes!

Here then is a basic difference between the matter-conversion systems of modern economics, and those of the biosphere. As we have seen, the biosphere knows no such thing as waste, and over the long-term course of evolution life has developed sound patterns of recycling. But in the short-term of human technological development, the human race has been too concerned with the benefits of products alone to balance the waste-matter budget. The whole pollution problem then is a consequence of the near-sightedness of human thinking, and the argument that pollution is a necessary partner of economic development is a false one. A highly satisfactory answer to the pollution question is available—planning and working within cyclic pathways that typify the biosphere.

Types and Effects of Waste Matter

The products of the waste-production arrow of human technology are described in Figure 24—2. There are three major categories of products, classified in terms of their biological consequences.

1. Wastes that are Normal Components of the Biosphere

These materials are discharged into the biosphere in abnormal quantity, either in one location, or over all. They are materials normally found in the biosphere however, and, in their normal distribution in both time and space, the biosphere has mechanisms to deal with them. They include human fecal wastes, nitrates and phosphates from sewage and fertilizers, sulfur dioxide, and many trace metals such as vanadium, and arsenic. Let us look at a few examples.

Pollutants like human sewage are released, have their effects, and

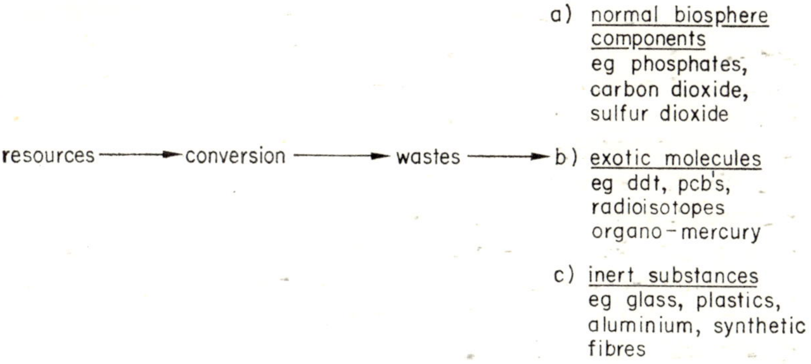

Figure 24--2 The One-Way Waste Production Arrow

become assimilated into the cycles of the biosphere. The effect tends to be highly localized and will be more pronounced with the addition of greater amounts of waste. We assume in this instance that the sewage contains only biological matter, and no industrial wastes.

The watershed into which the sewage is released contains a decomposer system, involved in the day-to-day business of recycling matter within this aquatic ecosystem. The capacity of these decomposers is limited by the amount of material supplied them by the ecostystem and, to a much lesser degree, by the amount of oxygen available during decomposition. Because the other organisms in this watershed—fish, amphibia, crayfish, for example—depend on adequate levels of dissolved oxygen (five to ten ppm) for their existence, it is essential that decomposition for the most part is not lower than these levels.

Human sewage is matter that can be easily oxidized by the decomposers of the watershed. It represents an artificial burden to the system, but if this waste is introduced in small quantities at a time when decomposer activity is well below the maximum, decomposition may be possible without lowering oxygen levels below the acceptable limit. However, as the volumes of discharge increase and as they catch the decomposer organisms at maximum activity, then the oxygen concentration of the water will begin to decrease and at certain points the river will become anaerobic. As this occurs, larger aquatic organisms will die off, particularly those intolerant of low oxygen levels. Organisms like the sludge worm and the midge—indicators of polluted waters—will begin to increase in number. Changes will begin to occur among the decomposer organisms as well. As oxygen levels decrease, aerobic decomposition will be replaced by anaerobic decomposition, whose products, instead of being sulfate and carbon dioxide, will consist of methane and hydrogen sulfide. These toxic materials will also affect the remaining life in the river, and should the wastes be great enough, the river will putrefy and die. If the wastes contain large amounts of phosphates and nitrates, then they will have further effects. When these nutrients enter a watershed they can accelerate plant growth. As the plants grow, they will fix carbon dioxide and produce oxygen. The excess oxygen will bubble out of the water. But when this excess of plant growth, which is usually algal, dies, it presents a greater-than-normal burden during decomposition. The oxygen content of the water thus decreases sharply. This effect, induced by the addition of fertilizing materials to watersheds, is termed *accelerated eutrophication.*

Ironically, a change of conditions that is detrimental to many forms of life has been induced by materials that in terms of their initial concentrations alone appear innocuous.

The situation described here is not always so simple. If the wastes are released into a quickly flowing river, they are carried downstream and may be deposited along the river over an extended area. The speed and volume of the water flow will both have effects upon the rate at which the sewage is metabolized and the degree to which water quality is altered.

Another example of this first category of waste materials is sulfur dioxide, a component of the atmosphere that is normally found only in trace amounts from volcanic activity. It is released by humans dur-

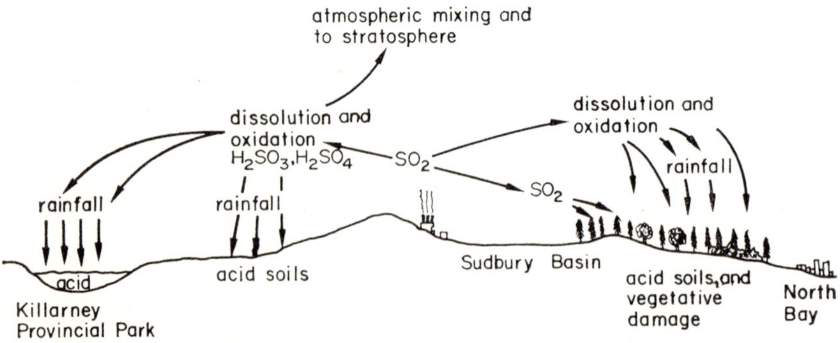

Figure 24—3 Sulfur Dioxide Pollution in the Sudbury Area

ing the burning of coal and other fossil fuels, and during the smelting of metals. As a contaminant it is spread through all phases of the environment—earth, air, and water. (See Figure 24—3.)

The Sudbury district of Ontario with its blackened rocks and denuded landscape presents an excellent example. Here, nickel sulfide ore is mined deep within the earth and smelted, with the production of nickel and great amounts of sulfur dioxide. Its primary effect is upon people, their possessions and plants. Lung diseases such as emphysema and bronchitis are aggravated. Traces of atmospheric sulfur dioxide will accelerate the corrosion rate of steel thirty times, of nickel twenty-five times, of zinc fifteen times. Stone, brick, and concrete all blacken and begin to crumble, rubber hardens, and textiles deteriorate. Coniferous trees and crops such as lettuce and alfalfa are very susceptible to sulfur dioxide—many farmers just outside Sudbury have given up. Depending upon the wind conditions the sulfur dioxide may be carried west or south-east of Sudbury. In either direction (almost sixty miles east to North Bay), forests have been damaged by sulfur dioxide. As it dissolves in the water vapour of the air, it forms a weak acid (H_2SO_3 or sulfurous acid), and in the presence of sunlight and oxygen this may be converted to H_2SO_4, or sulfuric acid. Both of these wash out of the air with rainfall and accumulate in the soils. Through time, the soils become steadily more acidic and more and more harmful to plant growth. In Killarney Provincial Park to the south-west of Sudbury and off the north shore of Georgian Bay, the lakes are quite shallow. Following rainfall—when the wind is from the Sudbury basin—the lakes turn suddenly very acid. Because of this, fish populations have collapsed. Furthermore, in the winter months the acid snow accumulates on the ice and the surrounding watersheds until spring. With spring run-off, this acid is very suddenly released into the waters. Spring is the spawning time for many fish, and fish eggs are very susceptible to changes in acidity.

By convection currents, sulfur dioxide is also carried into the stratosphere, with far-reaching effect. Acid rains have been observed in Scandinavia from smelters located in Germany, for example. On a global basis it has been estimated that human production of sulfur dioxide has increased the biogeochemical rate by 50 per cent annually.

Nitrogen oxides from car exhausts behave in a manner analagous to sulfur dioxide, entering the atmosphere from the burning of fossil fuels, causing great damage to plant life and coming down as nitrous and nitric acids.

2. Wastes that are Foreign to the Biosphere

These wastes are essentially foreign to biological mechanisms and there are few biological means existing to handle them. Most of them are acute biological poisons and tend to accumulate or concentrate in biological systems. They include mercury derivatives such as organo-mercurials; long-life insecticides and herbicides such as DDT and picloram; polychlorinated byphenyls (PCB's)—by-products of paint and plastics manufacture; and radio-isotopes such as strontium-90 from nuclear reactors and testing. They may include many unknowns among the thousands of new chemicals that we have introduced to the biosphere since the year 1900. They probably include a wide range of substances such as benzopyrenes (from car exhaust and coal-burning), dibutyl tin, and organo-lead chemicals.

As an example, let us consider the pollutant mercury. We have already discussed the use of organo-mercurials as biocidal agents in agriculture (Chapter 2—3). This use of mercury is only one of the many inputs to the biosphere. About a third of all mercury used today is employed as a catalyst* in the chlor-alkali process for the manufacture of sodium hydroxide (NaOH). Up to 70 per cent of the total mercury used in this manner may be lost in the wash waters and thus reach the environment. Deposits of metallic mercury have built up in this way in the bottom muds of the St. Clair River, from companies such as Dow and Wyandotte Chemical.

If the mercury were only to remain here, it might be a good while before its effects were felt (particularly since it is underwater, away from atmospheric oxygen and prevented from vaporizing). Bacteria in the bottom muds, however, can act upon mercury, methylating it and producing methylmercury (CH_3Hg). Methylmercury is an organo-mercurial compound, akin to the forms of mercury used as biocidal agents—for example, to protect pulped wood from fungal and bacterial growth during storage. One of the processes used in pulping wood employs caustic soda (sodium hydroxide). Very often, chlor-alkali plants will be constructed beside paper mills, and a river can receive mercury in the effluents of both plants. Other organo-mercurials are employed for seed dressings and to prevent bacterial growths in household products such as furniture polish and fabric-softeners. In total, 7 per cent of all the mercury used for human pur-

*Chemically a catalyst is a substance that accelerates or slows down a chemical reaction but undergoes no chemical change itself.

poses is employed in biocidal forms.

Mercury can thus spread through watersheds into which it is released. As bacterial action steadily converts the metallic mercury to organic mercury, this form of mercury biologically concentrates, reaching dangerous proportions in the upper trophic levels of the food chains. Here the characteristic effects of mercury poisoning—blindness, paralysis, and general deterioration of the nervous system—are displayed. Mercury contamination from pulp and paper led Sweden to curtail her inland and coastal fisheries. Mercury from a pulp mill at Prince Albert on the North Saskatchewan River led to the closing of fishing in Lake Winnipeg, 500 miles away. Because of mercury, fishing has been halted in Lake Erie, eastern Lake Ontario, and many rivers and lakes across Ontario.

3. Dead-End Materials

This class of waste materials consist of those inert to the forces of decomposition in the biosphere, and thus represent dead-ends, along which matter flows and stops. Including such wastes as plastics, hydrogenated cellulose, aluminum cans, glass bottles, and artificial fibres, these may contain material such as cellulose, needed by the biosphere in nutrient recycling, or they may contain material from outside the biosphere, such as nickel.

In many cases these pollutants are relatively harmless, simply because they cannot filter into biogeochemical cycles. (There is a possibility, however, that as some of these—plastics for instance—deteriorate slowly, they may release toxic molecules such as PCB's to the environment.) These wastes are aesthetically displeasing—often littering roadsides and public places—and some such as broken glass or tin cans may damage plant and large-animal life. Many of these wastes represent resources such as tin and fossil fuels and thus will be important factors in the problems of resource availability that will be discussed later in this chapter.

Many factors determine the ultimate effect of a pollutant upon the biosphere: the physical nature of the pollutant—gas, liquid, or solid; its solubility in water; its ability to codistil with water or its ability to adhere to particles of silt or clay. Will the pollutant travel with the wind, or move with groundwater? Will it be washed out of the atmosphere by rainfall? Is it acidic, or alkaline? Will it cause corrosion of surfaces, or, if it is a solid itself, will it slowly corrode?

Does it biologically concentrate? What are its physiological actions at various concentrations in microbes, animals, and plants? Does the material act as a biological fertilizer or stimulant? Does it act together with other materials that are present to cause *synergy*? In other words, two pollutants may act together in a manner that accentuates their total effect. Phenols, for example, are more toxic to trout when there is ammonia present, and vice versa. Phenols weaken the ability of trout to withstand abnormal levels of dissolved ammonia, thus effectively increasing the toxicity of the ammonia.

Pollution problems may often be complex. Industrial effluents may contain all three of the basic types of waste matter, and varied patterns of biological and physical forces may be transporting these about the environment and determining their ultimate effects. Pollution-abatement programmes must be geared to understanding the complex factors that are involved in any situation and must attempt to remove not only the most blatantly obvious pollutants, but also the more insidious varieties. Furthermore, pollution-abatement programmes must reconsider the whole concept of wastes and the linear production arrow of industry that produces them. Where the pollutant is in very large quantity, the most obvious answer is to find a use for it. Only by adopting the cyclic mechanisms of matter-flow in the biosphere will we truly be able to come to grips with the waste-matter problem.

The Economics of Waste-Matter Production

The release of waste matter into the biosphere has many consequences upon both the immediate human environment and the biosphere. These effects are hidden costs of the technology that produces them and have yet to be measured within the cost-benefit analysis of that process. How to measure these costs can be a difficult problem.

In many instances these costs may be borne by another sector of human industry. During 1949 the release of five billion gallons of pulp-and-paper wastes into the Puget Sound cost the highly susceptible salmon industry over \$24 million. In the United States, damage from air pollution amounts to a billion dollars annually, and in New York City alone it costs every household over \$500 a year in repairs and maintenance. Simply, the costs that are measurable by existing economic criteria such as these make pollution an expensive venture. Many of the costs arising from the release of waste matter, however, are not so easily measureable.

Consider, for example, the costs to human health and welfare, from respiratory ailments and DDT-induced cancer. Consider the youngster in Richmond, B.C., who fell into an open streetside sewer and caught strep throat, from which it took ten weeks to recover. (Collecting human sewage in this manner is not uncommon in North America.) Falling in again, she became infected with eight different vaginal diseases that required another six weeks for recovery. Consider the damages to beaches and other recreational facilities from masses of stinking algae and contamination by sewage effluents, from oil-spills and industrial effluent, and from litter. All of these are costs paid by the general public and particularly the underprivileged sectors of modern society, who have neither the money nor facilities to move farther abroad and escape these conditions. In many ways these costs are reflected in higher health bills. In others, however, such as the loss of freedoms and the deteriorating quality of life, they have not been given a dollar-and-cents value. Yet there are further costs that are even more difficult to determine.

These are the effects of waste matter upon the biosphere—upon the common air, earth, water, and living organisms that are essential to all of us, but that do not fall into any of the traditional spheres of human economics. How do we measure, for example, the costs of DDT, expressed in the near-extinction of birds such as the loon and the osprey? The transformation of thousands of square miles of countryside to barren rock by sulfur dioxide is certainly a thing of great consequence, but when the land has no particular human value, in terms of recreation, forestry, or agriculture, how do we measure the costs? What of the damage by pulp-and-paper mills to rivers like the Abitibi that are hundreds of miles from dense human settlements, how are these damages assessed? Yet all of these commons serve vital ecological roles. These roles are often long term and beyond the twenty-year scope of economic thinking, but they are nevertheless important ones in the future of humanity and the biosphere. For a long time, much of the biosphere has remained undefended and unprotected, considered as open for simple exploitation by those with the resources and power enabling them to do so.

There are two distinct, new outlooks that must be applied to the economic systems of today and the waste matter that they produce. First of all, an attempt must be made to evaluate the costs of pollution on the human and ecological environment. The costs accrued by human society alone, through ill-health, through damage to property, to other sectors of the economy, and to recreational and other common facilities are very high. Indeed it is very likely that together they would at least surpass the costs of installing and operating pollution-control devices.

Secondly, we must attempt to weigh long-term factors and values into our economic systems. Perhaps this situation is best exemplified by mercury pollution in Lake Erie. Directly, costs are incurred by the fishing industry through loss of sales. In the long term however, mercury is an insidious biocide that will affect this ecosystem for many years to come and will act in concert with other pollutants that are now present to accelerate the decline of Lake Erie. Beyond the long-term value of Lake Erie as an ecological component of the biosphere, there is a value to be placed upon mercury as a non-renewable resource. Mercury is a rare and a valuable metal, essential in many of the processes of human industry. Although on a one-year basis there is no shortage of mercury, the cost of this wastefulness will be paid for by humanity in the next twenty years with a severe mercury shortage. As we shall see in the next section, there is a finite supply of the non-renewable resources that are necessary for human industry, and present-day use must be evaluated in terms of the human future.

The Size of the Arrow

Not only are the waste materials at the end of the technological arrow of major environmental significance, so also are the raw materials that feed into it. The quantities of these resources that are used determine the size of the arrow and the amount of its wastes. And the global availability of all of these resources places a series of limits on the technological activity of the human race. In many ways human consumption of these resources is already meeting the limits that exist, and yet economic systems fail to recognize these limits. In this section we shall consider the availability of non-renewable resources and the limits that they impose. Following this, we shall deal with some of the trends and patterns of resource consumption that exist today, and what they augur for the future. Let us start with a look at the fundamental nature of non-renewable resources.

Non-renewable resources are obtained from the crust and oceans of the planet. Their availability in the short term is limited by human ability to reach and to utilize them. Over hundreds of thousands to millions of years these resources may be geologically replenished, but it is highly doubtful at this point that we should use these as criteria in planning resource usage for the next 500 years! Once used, then, unless they are recycled, these resources are irreplaceable. They include metallic ores such as bauxite (aluminum), chalcopyrite (copper), and magnetite (iron), fossil fuels such as coal, natural gas, and petroleum, and materials such as phosphate rock, limestone, gravel, asbestos, potash, and rock salt.

Distribution of Non-renewable Resources

Rocks are formed of components termed *minerals*. On closely examining a piece of granite for example, one might find it made up of a very close-packed mass of thousands of crystals of feldspar, quartz, and hornblende. Each of these types of crystal is a mineral.

Chemically minerals are quite homogenous. Each type of mineral consists of specific proportions of the atoms of certain elements arranged together in specific orientations. Both the orientation and the ratios of each of these types of atoms are specific for a mineral. Both of the iron minerals goethite and lepidocrocite have the same chemical constituency: two iron atoms to three oxygen atoms to one water molecule ($Fe_2O_3H_2O$). The difference between the two, however, is a result of a different arrangement of the components. On the other hand, chalcopyrite and bornite both contain the same type of atoms but in distinctively different ratios. Chalcopyrite is $CuFeS_2$, whereas bornite is Cu_5FeS_4. Slight changes in the ratio of components, the presence of trace amounts of uncommon elements such as arsenic, uranium, or silver, and differences in the conditions of formation all lead to different types of minerals.

Fundamentally, the rock of which the earth is formed is composed of minerals, and the composition of these varies from area to area. Each mineral is composed of the atoms of various elements in certain proportions. On a global basis, silicon and oxygen are the elements most commonly found, followed by iron and aluminum. By comparison, nearly all of the other elements are extremely rare. There exist distinct differences in the way that these elements are distributed among the rocks of the earth. Figure 24—4 shows two possible modes of distribution.

The first mode is termed a *continuous* one (Figure 24—4A) and is applicable to deposits of iron and aluminum and a few other minerals including porphry copper ores. First of all, the element is found in the average concentration of normal rock. These are the poorest deposits of the element one can find, but the most extensive as well. At the other extreme there are highly localized and very, very rich deposits of this element. These are the deposits that are usually exploited at the first opportunity. In between these two extremes are a whole range of deposits, each of gradually decreasing quality but each of gradually increasing size. If, for example, the richest ore in this pattern contained 40 per cent iron, of which on the surface of the earth there were 't' tons, then of 35 per cent iron there might be 2t tons, of 30 per cent iron 4t tons, of 20 per cent iron 16t tons, and so on. A distribution like this one has distinct advantages. When the richest ores are totally exhausted, one can always turn to the next richest, which are there in slightly greater amounts and can thus compensate for the decrease in quality. Technologically speaking, this is ideal, because a slow increase in efficiency over time can help to adjust for the gradual drop in quality.

This ideal situation, however, does not predominate. Most of the other elements are distributed in the opposite fashion. Because rich deposits, particularly metallic ones, but also salt, sulfur, and fossil-fuel deposits result from highly localized geological activity—the bed of an ancient ocean, intrusions of molten lava into the bed rock, and processes of selective erosion, weathering, and precipitation—then

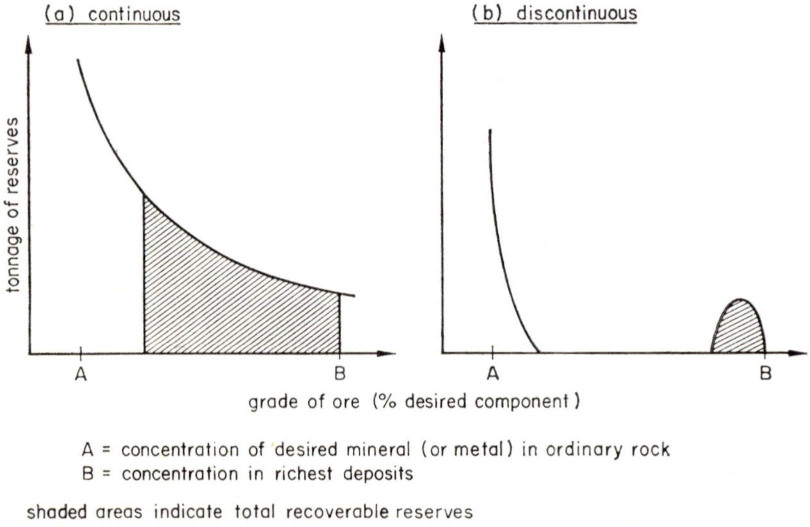

Figure 24—4 Patterns of Mineral Distribution

deposits of all but the most common elements tend to be highly localized. They are, therefore, found either in the minute concentrations of normal bedrock or in the very high concentrations of localized deposits, but not in a continuous range of concentrations in between. Elements like helium, sulfur, mercury, silver, the rare earth metals, uranium, chromium, tin, copper, nickel, and minerals such as oil, sulfates, and potash are usually found in this way. This sort of distribution is termed *discontinuous* and is illustrated in Figure 24—4B. Once we have exhausted the richest deposits of these minerals we are faced with almost no other alternatives but to mine them from common rock.

How then are the reserves of these minerals determined? First, let us consider what the term 'reserves' means. Let us suppose that the copper ore chalcopyrite occurs in a number of known locations. In each of these locations, geological surveys have partially delineated the extent to which chalcopyrite occurs. The sum total of the chalcopyrite in these definitely known amounts constitutes our *known* reserves of chalcopyrite. They are thought to extend farther, beyond these areas, because the rock formations appear similar, but this has not been conclusively proven. The total of these areas constitutes the *inferred* reserves of chalcopyrite. But apart from these specific deposits, there are likely other deposits of which we have no knowledge. We know the types of rock structure in which we can expect to find chalcopyrite, and since the locations and extents of these structures are known, it is therefore possible to postulate the amount of yet-undiscovered chalcopyrite. These are the *speculated* reserves. Altogether, these three classes constitute the *ultimately recoverable* reserves of this copper ore.

Modern geological techniques are being employed to document and exploit accessible reserves of minerals. Some of these techniques are carried out fairly rapidly and on a large scale. Aerial magnetic surveys, for example, can determine magnetic changes and patterns in rock structures over broad areas—possible locations for ore bodies. Some plant species grow well in the presence of specific minerals, and prospecting can involve a general survey of vegetation in order to spot peculiarities that may signify a mineral deposit. Seismic studies using shock waves is another method. On the other hand, locating the actual ore body may be a more difficult process. For example, although it is fairly easy to predict the type of geological formation in which oil deposits occur, repetitive and tedious drilling must be carried out in order to reach a single pool.

What of the minerals located on the floors of the oceans and dissolved in sea-water? Surely these must also be included in world reserves, for even though we may not be able to exploit them now, future technology may provide us with the means. The ocean floors, geologically speaking, appear to be very young. According to the theories of continental drift, from the hot core of the earth outwards convection currents of molten rock are carrying the continents farther apart, with the formation of new sea floors between them. As the continents move forward, the old ocean floors sink beneath their front edges.

While the rock of the continents is relatively old, the rock of the sea floors has been recently formed. The continental rock has felt the strain of the earth's crust as it cooled, and the strain of continental drift; it has been subjected to the forces of mountain building and erosion for billions of years. Within the cracks and fissures of the continental rock, the volcanic intrusions and all of the other geological processes responsible for mineral deposition have occurred. It is thus for good geophysical and geochemical reasons that rich mineral deposits are more likely to be found in the continental areas than in the newly formed ocean floors.

We may find some resources on the floors of the ocean. In some locations, at depths between 3,000 and 6,000 metres, there are 'manganese nodules', rich in manganese and also in iron, nickel, cobalt and copper. Some future technology may perhaps be able to harvest these, with great energy expenditure. To date, however, no major ore body has been located on the ocean floors.

The continental shelves (the shallow sea floors near continents) are a different matter. These are extensions of the land masses and represent potential sources of many of the minerals located on the continental areas. These include deposits of phosphates, oil, potash, manganese, tin, diamonds, and gold, some of which are already being exploited.

Sea water itself contains many minerals (Table 24—2), but the concentration of most of these elements prohibits their recovery. Take zinc for example. At current market prices, processing a million gallons of sea water would provide only 1·3 cents worth of zinc. To break even after covering operating costs and salaries, an industry would have to process a minimum nine trillion gallons of sea water and produce 400 tons of zinc. Yet today the United States consumes over 125,000 tons of zinc annually. Except, then, for the most common elements in sea water—magnesium, bromine, sodium, sulfur, potassium, and iodine—and possibly fluorine, strontium, and boron—it seems unlikely that sea water will approach our demands in either an efficient or an economic way.

Table 24—2 Concentrations of Some Elements in Sea Water*

Element	Concentration (ppm)†	Recoverable value ($ per million gal.)
Chlorine	16600	-
Sodium	9200	-
Magnesium	1180	4130
Sulfur	775	101
Bromine	570	190
Strontium	7	-
Boron	4	-
Rubidium	0·1	125
Phosphorus	0·06	0·08 (as $CaHPO_4$)
Iodine	0·05	1
Zinc	0·009	0·013
Iron	0·009	0·001 (as Fe_2O_3)
Tin	0·003	0·05
Copper	0·003	0·01
Nickel	0·002	0·02
Silver	0·0003	0·02
Lead	0·00003	0·00004
Mercury	0·00003	0·002
Gold	0·000004	0·02

* Adapted from *Resources and Man*, W. H. Freeman and Co., 1969.
† By weight.

For our mineral supplies we are therefore largely limited to the continental areas of this planet. How accurately can we estimate the reserves of minerals available to us? Much of the continental area of Earth has been surveyed, and many major and minor mineral deposits and their subsidiaries have been located. To a large degree we know the geographical extent to which these deposits occur. Beyond this we have a good idea of rock formations and the likelihood of as-yet-undetermined mineral deposits. Various estimates of world mineral supplies have been made, and some early ones are estimated by Landsberg and given in Table 24—3. Using 1970 consumption rates in the United States for each of these resources, it has proven possible to place a rough life-time on these reserves, should the whole world of 1970 consume at a similar rate. Apparently only four of these materials would last more than ten years, and only iron beyond half a century—assuming that consumption rates remained constant at these levels, which is not a good assumption.

Table 24—3 World Mineral Reserves and Potential World Demands

Metal	World reserves (millions of tons)	U.S. 1970 consumption (millions of tons)	Reserve* lifetimes
Iron	129,000	100	65
Aluminum	3,000	4	43
Copper	270	2·5	6
Manganese	450	2	13
Chromium	500	1	28
Nickel	16	0·3	3
Tungsten	1·4	0·012	7
Molybdenum	3	0·06	3
Lead	80	1	4
Zinc	140	2	4
Tin	5	0·28	4

* Assuming that a 1970 world population (3·5 billion) consumes resources at a per-capita 1970 U.S. rate (U.S.=0·2 billion people).

The conclusions drawn from this table are supported by the predictions of many scientists today. Most notable is the report *Resources and Man,* published in 1969 by a committee of the U.S. National Academy of Sciences and National Research Council set up to study the relationships between human population and resources. The committee predicted shortages, in the next thirty years, of mercury, silver, tungsten, copper, lead, gold, zinc, beryllium, tantalum, the rare earth metals, and helium. For mercury alone, the report notes that if world production over the next twenty years were to rise only a half of predicted demand in the United States, more mercury would have to be mined than is contained in the estimated world reserve.

In the face of impending mineral shortages, can we turn to technology for an answer? The technology being used to locate, extract, and process minerals is continually improving. Can it provide us with substitutes for minerals that are running short, and if not a substitute, can it provide a cheap and easy way of producing the necessary material from common rock, or sea-water? The great benefits from technology in recent years have led people to believe that there is no limit to what it can accomplish. Let us consider some of these claims.

There appear to be very real points at which the immediate benefits of technology are overcome by the disadvantages. For example, what will happen when we have to mine a metal such as lead from common rock? Greater amounts of the rock will be needed, and greater amounts of energy to mine it, crush it, transport and process it. Breaking up the rock in order to process it, increases its size about twofold. Processing requires large amounts of chemicals for treatment. Thus, to produce the same amount of lead that was previously obtained from the richest ores requires far more energy and presents massive waste-disposal problems. Certainly, in this light, the disadvantages may outweigh the benefits. To go through this process for all of our mineral diet would be ludicrous. As proof, let usconsider some recent trends in the processing of minerals. Figure 24—5 shows the amount of energy put

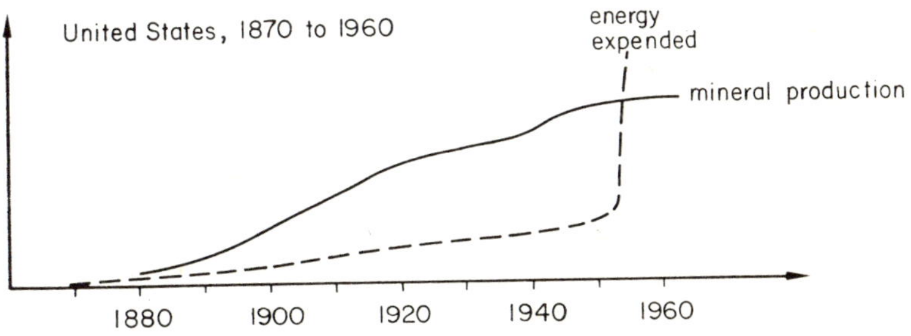

Figure 24—5 Energy Consumption in Mining

into the mining of ores in the United States, in terms of the amount of metal produced. Although between 1950 and 1960 metal production actually levelled off (imports were supplanting home supplies), the amount of energy consumed actually tripled. In other words, the use of poorer ores and the transport of larger amounts of these have led to a sharp rise in energy expenditures. In the U.S. copper industry the increased costs that arise from the exploitation of lower-grade ores are barely allowing the industry to break even.

If technology cannot keep up to impending shortages, is it likely to find substitutes for essential uses of scarce metals? To a certain extent this is already happening. Aluminum is becoming a light-weight substitute for many uses of iron, copper, and silver. Plastics are replacing many of the uses previously made of metals. Yet silver is still essential to photography, and there is no apparent alternative for it. No superior alternative to copper has been found for use in electric circuitry and communications systems. Helium as an inert gas, mercury as a liquid metal, and uranium as a fissionable radioactive element have properties that are almost unique. Besides this, technology is finding still newer uses for these scarce metals.

In conclusion then, the ability of technology to extend the availability of non-renewable resources is limited. It is in some cases easing the situation, but it is definitely not the provider of boundless resources. All of the reserves of non-renewable resources on this planet are exhaustible and, in many cases, we are in danger of doing so in the next thirty years. This crisis is being precipitated by very high consumption rates, and these rates are linked to a number of factors within the various sectors of the human world. Let us consider some of them.

Patterns and Politics of Resource Consumption

There appear to be three very fundamental factors that interrelate in the generation of resource demands.

First of all, population growth is directly linked to the consumption of resources, assuming that at least a constant per-capita level of usage is maintained. Table 24—3 has already illustrated some of the basic limits that exist on world-population size, in terms of non-renewable resources.

Secondly and thirdly there are the factors of economic growth and economic development. Both are expressed as an increasing Gross National Product or G.N.P. Because money is a symbol for goods and services exchanged, both are linked to an increasing demand on world resources. Despite this common similarity, there are important differences between these two phenomenon, differences that are essentially cultural.

Economic development is an early form of economic growth that is employed by the developing nations in an attempt to improve fundamental living conditions through better health and welfare services and modern educational facilities. This requires an increased consumption of resources. But Table 24—4 illustrates what would happen should the people of the developing world attempt to reach a level of per-capita material prosperity with the developed world. Here the background of metal behind a citizen of the developed world—in vehicles, communications systems, buildings, machinery, appliances, homes, schools, hospitals, etc. are used to calculate the proportion of the estimated world reserves necessary to bring all of today's population to this level.

Table 24—4 Resources for World Development

Metal	Tons to bring 1970 world to U.S. level	Obtainable reserves	% of World reserves required
Iron	30 billion	129 billion	23
Lead	500 million	80 million	625
Zinc	330 million	140 million	235
Manganese	180 million	450 million	40*
Molybdenum	50 million	3 million	600†
Tin	50 million	5 million	1000

* 12lbs. of manganese per ton of steel, to convert all of iron to steel.

†1·2lbs. of molybdenum per ton of steel, to convert all of iron to steel.

Although there might be enough iron in the world's reserves, there would hardly be sufficient tin, lead, zinc, or molybdenum. And should the developing nations wish to emulate Western society they might be doomed to failure. The problem lies partially in the high rates of consumption in the developed world. Already a majority of world consumption takes place in North America, where 6 per cent of the world's people consume 40 to 50 per cent of the world's resources consumed every year. Total consumption is to double by the year 1990, while the North American population, because of slower growth-rates, is expected to drop to 5 per cent of the world's total.

These high consumption rates in North America, which are typical also of the rest of the developed world—Japan, Europe, the Soviet nations, and Australia—are indicative of nations in advanced stages of economic development. This stage of overdevelopment is termed economic growth.

Economic growth could be defined as growth for the sake of growth, popularly termed the ideology of a cancer cell. Concomitant with an increasing G.N.P. are rising material conventional standards of living and a falling environmental quality. It appears, in the Western world at least, that a steadily increasing G.N.P. is tied directly to the stability of the socio-political system and is equated with satisfactory economic and social conditions. Should the G.N.P. cease to rise, the system would be faced with inflation and disorder. Yet in the Western world, an increasing G.N.P. is no longer linked to the improvement of basic living conditions in the same way as economic development in the underdeveloped world is. (This is not to say that the developed nations have wiped out poverty—this is far from the case.) Economic growth appears to be, in part, a phenomenon by which the multi-national corporations maintain their existence, by increasing gains through profits and through empire-building by means of branch plants about the world. Economic growth is a phenomenon that, although providing certain benefits and improvements in living standards to some people of the developed world, also results in disadvantages such as pollution, loss of freedom, and cultural manipulation. The major benefits of economic growth are directed at the shareholders and the management of the corporations and the financial world. To keep the G.N.P. growing, the multi-national corporations have resorted to a fistful of wonderful and devious means.

Products are sold on as wide a basis as possible, and advertising is designed to increase consumption. Advertising is employed to convince the consumer that he needs the product, whether or not he really does. It attempts to do this by exploiting the individual discontents of the people to whom it is selling. It suggests in many cases that material goods will replace and satisfy social ills and personal dissatisfactions, indeed even to the extent of causing an overnight transformation in personality!

The products that are sold range from worthwhile to worthless. They are sold not because they aid the customer—that is an additional advantage if it happens to occur—but because they increase sales and the G.N.P. Consider, for example, the dubiously 'essential' appliances from electric toothbrushes to electric knives that flooded the markets in the late sixties. Products are often designed in such a manner that they deteriorate or break down after a short period of use. Cars rust away after several years, radios and record-players burn out, and clothing lasts only a season's wear. In other words, after products have been used for a short time, consumers are forced to replace them with identical ones, and this contributes to an increase in economic productivity and an increase in the G.N.P. Other products are advertised and sold that are mere gimmicks and serve only to waste energy and resources, for example electric nail-polish dryers and executive playpens.

The strong obsession of our Western system with economic growth is one of the most serious problems that we face. Not only is it contributing to a steadily deteriorating environment, but in the light of world mineral reserves and the needs of the underdeveloped countries to achieve a satisfactory standard of living (let alone two cars per household and three washrooms per five people), it is likely to create major political tensions in the next thirty years as it attempts to get the resources necessary to continue this growth. Furthermore, there are also many social problems arising from an obsession with material welfare and a way of life that attempts to satisfy human social and cultural needs in terms of material luxury.

Summary

There are limits on the ability of the earth to supply us with non-renewable mineral resources, simply because the earth is finite and for the most part these resources occur in specific locations of finite extent. Technology, it appears, is very much a panacea for the refusal of man to accept these limits. Although technology will certainly help to ameliorate the situation of resource shortage, both by finding substitutes for scarce resources and by devising more efficient methods of utilizing lower-grade ore bodies, technology itself has limits, and we must not place faith in it alone. Newer and better 'pollution- and problem-free' technologies are a dream. They are likely to present greater problems than we have now—problems that we are as yet not able to perceive fully. They are certainly not the answer to the present crisis. And perhaps we have been guilty of using our delusions of grander technology as an excuse for present inaction (or actions).

The industrially developed nations of the world are using a disparate proportion of the world's non-renewable resources, and should today's world population attempt to reach equality on a basis of per capita consumption with the presently developed nations, then we would likely have severe problems in providing them. The developed nations of the world are manipulating the available resources to feed their own needs, through political, financial, and military manoeuvring that gives little thought to the future needs of either the developing nations, of humankind as a whole, or of the biosphere.

We must return to the first base of production economics and study the sensitivity of the methods of cost-benefit analysis that are being employed. Do they fully account for all of the environmental and social costs of the product being produced, and do they place the use of resources and energy in a short- or a long-term perspective? In the general picture, we must search for new directions in which to bend the waste-production arrow. One of these is recycling. We must also fully re-examine the basis of our economic philosophy of growth. In the Western world it is insane to think that we can continue along an exponentially increasing pathway of consumption without crashing headlong into Earth's limits and our own destruction.

CHAPTER 2—5

Energy : Fuel for the Fire

Energy, it will be recalled, cascades through the biosphere in a manner that drives the biogeochemical cycles. Whereas matter is recycled following use by the processes of life, energy is irrecoverably lost as waste heat. Almost all of the energy used by the biosphere ultimately comes from Sun. Energy is also essential for human technology—for example, to extract ores and to process these into metals, to transport these metals and to convert them further into products. In the case of products such as the automobile and the light bulb, energy is also needed to operate them.

If humankind were to rely solely upon the flow of energy within biological systems, our modern demands would be hardly satisfied. Traditionally we have employed wood for cooking, heating, and smelting, but this resource is highly limited and bulky to transport. Apart from wood or peat there have been few biological materials that could supply us with copious amounts of energy. Biological systems can produce both light (the firefly) and electricity (the electric eel), but this production is discontinuous and impossible to harness. For the most part, energy flowing within living organisms moves at the level of the cell in a series of relay systems that are geared to one purpose only—the continuity of biological life. Some of this can be harnessed directly by using the living organisms as a whole, for example, the horse or ox. Others, such as yeasts, are used for highly specific food processes such as wine-making.

Civilization has, therefore, turned to other sources of energy, and, in the developed world at least, these have to a very large degree supplanted biological energy sources such as the horse. Early alternatives included wind and water power; coal, gas, and oil were later introduced. More recently we have been turning to nuclear fission. Energy demands are dependent upon the magnitude of industry and are linked, like the consumption of material resources, to population growth and economic growth and development. As each of these increases, the reserves of energy-rich materials decrease and the effects of energy-generation upon the biosphere become more intense. With the use of energy, like that of other renewable and non-renewable resources, the human animal is up against the fundamental limits imposed by the biosphere. Let us consider how energy is generated for human needs, and what the effects of this are. Later we shall consider the limits imposed by the biosphere and the planet.

Generating Energy

A fundamental and very useful property of energy is its ability to change from one form to another. Coal, wood, oil, or natural gas may be burned to produce heat energy for cooking or heating. Gasoline and other fuels may be burned to produce mechanical energy, which can be employed to drive vehicles and machinery. By far the most convenient form of energy is electricity. Electrical energy can be easily transported, and then transformed into many forms including heat, light, mechanical, and chemical energy. Electrical energy may be generated by a wide range of means, but, nearly always, kinetic energy is used to rapidly drive turbines, which then generate electrical power. In hydroelectric power, the kinetic energy of falling water (gravity) is used to drive the turbines. In geothermal power, hot springs and geysers are tapped and the pressurized steam from these is employed to drive the turbines. In thermal-electric power, fossil fuels are burned to heat the steam, and in nuclear power, nuclear fuels such as uranium-235 are burned to produce the steam. All of these processes have distinctive problems and effects within the biosphere.

1. Fossil Fuels

All of the processes for obtaining energy from fossil fuels produce carbon dioxide and use up atmospheric oxygen. They do not, however, constitute a major threat to the world's oxygen supply. Although the earth's crust contains over forty-five times the amount of fixed carbon necessary to account for the free atmospheric oxygen today, the vast majority of this carbon is highly dispersed. The fossil fuels probably represent less than 0·1 per cent of this total, and the oxygen concentration of the atmosphere appears to be stable for a good while into the future, at least. Carbon dioxide, however, is a different matter. Whereas oxygen constitutes about 20 per cent of our atmosphere, carbon dioxide is only about 0·03 per cent, and the addition of carbon dioxide from the burning of fossil fuels alters this appreciably. This point will be dealt with a little later.

Besides carbon dioxide, the combustion of fossil fuels can produce a wide array of minor products that plague urban areas. These include particular matter, such as fly ash, aromatic and carcinogenic hydrocarbons such as benzopyrenes (these first two from coal in particular), sulfur dioxide, nitrogen oxides, arsenic, fluorides, and metals like vanadium, cadmium, and mercury. The thermal-electric generation of power is of major importance in industrial centres across the United States, Canada, Europe, the U.S.S.R., and Japan, and provides well over 90 per cent of the world electrical supply. It has also been classed as one of the major sources of pollution in the urban environment, particularly of sulfur dioxide.

Coal is one of the most widely used fuels for generating thermoelectric power. But of all the fossil fuels, coal is the least ideal. Not only is it more difficult to obtain and transport than any of the others, but during its mining widespread environmental disruptions occur, and during its use it is a major pollution source. Coal may be mined deep in the ground over large areas, often causing shifts in rock structure and minor earthquakes. In large areas of Pennsylvania, Ohio, Kentucky, Virginia, and Tennessee, extensive coal deposits follow the contours of the hills at a short distance below the surface. Often these deposits are mined by removing the trees and topsoil and simply excavating. Moving along the hills, strip-mining has destroyed thousands of acres of scenic land, accelerated erosion, destroyed the vegetative cover and the water table, and caused cave-ins and landslides. In abandoned mines, a bacterial process is at work on some of the coal remains to produce highly acidic wastes, which seep outwards into lakes and streams. This has destroyed thousands of miles of watershed. Similar effects have been noted in Wales and the English Midlands.

2. *Hydro-Electric Power*

Building dams can have widespread ecological effects. Above and below the site, changes in water temperature, upsets in nutrient cycling, changes in aquatic and associated life-forms, and erosion can occur. In tropical areas, eutrophication and intensive plant growth in the reservoirs behind the dam often occur, as well as drastic changes in water table both upstream and down. Dry areas have become wet and invaded by mosquitoes, and wet areas have become dry and invaded by scorpions. Streams below the dam have readily become clogged with silt and the parasite schistosomiasis has often set in, for example with the new Aswan Dam in Egypt. Because of erosion and the slower movement of water, dams silt up and eventually become useless. Sometimes dams become silted full in a few decades, elsewhere they may last centuries.

3. Nuclear Energy

Every atom consists of a central nucleus surrounded by a cloud of electrons. The nucleus appears to be made up of protons and neutrons,* the total number of which largely determines the atomic weight of the atom. The most common form of carbon contains six protons and six neutrons and is referred to as carbon-12. While the neutrons contribute equally to the nuclear mass, the number of protons determines the chemical properties of the element. Were there five protons instead of six, then the atom would be one of the element boron; were there seven, it would be a nitrogen atom.

Although carbon-12 is the most commonly found variety of carbon in nature, other forms do exist. These forms are called *isotopes* and differ from carbon-12 in the number of neutrons in their nuclei. A combination of 6 protons and 6 neutrons is more stable than 6 protons and 7 neutrons (in carbon-13) and 6 protons and 8 neutrons (in carbon-14). These unstable isotopes break down slowly, releasing energy and more stable isotopes. They are thus *radioactive* and are termed *radio-isotopes* of carbon. Carbon-14 for example breaks down to give:

$$\text{Carbon-14} \longrightarrow \text{nitrogen-14} + \text{electron} + \text{neutrino}$$

The half-life for carbon-14 is 5,000 years; in other words, after that period of time one-half of the carbon-14 in a sample has decomposed. The rates of decay of radioisotopes cover a large range of values; calcium-39 has a half-life of 0·7 seconds, whereas calcium-48 has a half-life of over twenty million billion years.

Among all of the atoms making up the earth, there exist a small proportion that are radioactive. These are dispersed and diluted by large quantities of stable or 'cold' matter. A large proportion of the

*A wide range of sub-atomic particles have been postulated and/or detected in the last fifty years. The most common are those making up the atom, the proton (with a positive electrical charge), the neutron (electrically neutral), and the electron (negatively charged). In terms of mass, the proton and the neutron are about equivalent, and the electron is about one-two-thousandth of this mass. Other particles that have been discovered include the muon, the meson, and the neutrino.

elements of high atomic weight are unstable isotopes; indeed, some of these elements have no known stable forms and are all radioactive. These include thorium, radium, and uranium. Of all of the naturally occurring forms of these elements, however, only uranium-235 exhibits an unusual and interesting property. Bombarding uranium with a stream of neutrons causes it to disintegrate rapidly, in a number of possible ways, the most common being:

$$\text{uranium-235} + \text{neutron} \longrightarrow \text{barium-144} + \text{krypton-85} + 3 \text{ neutrons} + \text{energy}$$

Barium-144 and krypton-89 are both unstable and disintegrate further. In the decay process, three neutrons are emitted which, in turn, can cause the decay of other uranium-235 atoms. This process is self-perpetuating, and uranium-235 is thus termed a *fissionable* or a *fertile isotope*. By purifying this element and bringing quantities of it together in a controlled manner, one can induce a chain reaction. The large quantities of heat energy that are produced can be employed to generate electricity, and this is the basis of operation of the *fission reactor*.

Radioactive materials can be extremely hazardous. The radiation that they emit is often of a highly energetic and penetrating form termed *gamma radiation*. This, along with the sub-atomic particles that are emitted at high speeds, is a powerful disorienting force on the molecules of life, causing disruption of cellular function, mutations, and even cancers. Furthermore, radioactive elements may be concentrated manifold from the environment by living organisms and magnified through the food chains. Fixed inside the molecular structures of a cell, they continue to disintegrate, releasing energy and causing extensive damage within the cellular environment. Radioisotopes such as iodine-131 and strontium-90 can be particularly dangerous, as they will accumulate in one specific area of the body such as the thyroid or the bones. (Strontium is chemically very like calcium, a major constituent of bones.)

From an environmental point of view, no methods that are entirely satisfactory have been devised for the disposal of radioactive wastes. In the U.S. some highly radioactive wastes have been injected into shale deposits at depths of 700 to 1,000 feet, but it is questionable that even here they are completely away from the water table. Should geological shifts occur and fractures break up these shales, some of these wastes may very well be released to the biosphere. Another practice involves sealing the wastes in abandoned salt mines where they become trapped in a cloak of rock salt. Very recently these mines have been shown to be permeable to the water table. Low-level wastes have been treated with greater abandon. Great Britain dumps all of her low-level wastes into the oceans. In the United States they are often buried in shallow (ten to fifteen feet deep) trenches—well within reach of ground-water circulation. As a result of seepage, contamination of the water table has revealed itself in the biological concentration of radioisotopes such as zinc-65. The Ontario Hydro Commission stores most of its low-level wastes until decontaminated, but some very small amounts are deliberately released in the effluent at what are supposedly safe levels. Gaseous wastes from nuclear reactors are very difficult to trap. They include tritium (hydrogen-3) and krypton-85; no attempts are being made to trap the latter. Krypton-85 has a half-life of ten years, and it has been estimated that should nuclear plants go into operation as planned, then levels of krypton-85 in the atmosphere would reach dangerous proportions by the year 2000. In that year, the background radiation received by every individual will be raised to two-thirds of the maximum permissible by today's standards.

Radioactive wastes may also escape from the reactor itself. In the United States, the heat energy is carried away by steam flowing over the reactor. Separating the reactor and the steam is a closed jacket of light or normal water that conducts the heat from the reactor core to the steam. As a consequence of this system, fuel rods can only be replaced when the reactor has been shut down. Power utilities, therefore, drive their reactors as long as possible before changing fuel supplies. Corrosion of the reactor vessel and a build-up of radioactive material is therefore inevitable, leading to possible contamination of the cooling waters.

In Canada, the CANDU system appears to be much safer. Heat transfers are accomplished with heavy water (water containing deuterium or hydrogen-2 instead of hydrogen). Heavy water is a good absorbent or moderator of neutrons, and Canadian reactors can, therefore, be better regulated and the fuel rods changed while the reactor is in operation. Furthermore, the Canadian reactors require a less purified form of uranium, consequently there is less chance of contamination during the manufacture of fuel rods. It is interesting to note that the CANDU system is already facing stiff competition from the inferior American systems, and this pressure may push it out of existence.

In the biosphere there is a constant background of radiation from the decay of radioactive materials and from Sun. It is to these levels that life has adapted, and it is certainly these levels that were important in the continuous rate of background mutation that was essential to evolution. Increasing these rates is, therefore, a thing of great controversy. In the United States, the Atomic Energy Commission (A.E.C.) claims that 170 millirads* exposure per year is the total maximum allowed for an individual of the general public. But Dr E. Sternglass, who has done statistical correlations of mutation rates and malformed births with fall-out levels from nuclear testing, argues that this is too high. This level has also been challenged by Dr A. R. Tamplin and J. W. Gofman (of the Lawrence Radiation Laboratory at the University of California). They contend that there are no minimally justified levels except for the background geological ones —a concept long held by many geneticists. The average geophysical background radiation received by an individual each year averages about 50 millirads, less than a third of the allowable maximum. (For persons working with radioactive materials, a maximum of 1,700 millirads is permitted.) All of its critics have received the heavy censure of the A.E.C. for daring to question the laws set by this body that is its own policing agency.

It is not so much the present dangers of nuclear wastes that are a cause of great concern, but the projected wastes, should the nuclear-power industry expand its production as expected. By 1980, nuclear power could provide 10 per cent of the electricity of the developed world. According to *Fortune* magazine, between 1980 and 1985 in the United States alone, in order to keep up with demand, a new 1,000 megawatt reactor will have to go into operation every two *weeks*. And, as Paul Ehrlich notes, in the year 2000 in the United States at any one time, at least 3,000 six-ton trucks will be carrying reactor wastes to disposal sites. Not only will this present a serious disposal problem, with a much greater chance of environmental contamination, but there is the very real possibility of one or more of these trucks being in an accident. It is noteworthy also that no technology exists today to dismantle obsolete reactors. Will these clutter the countryside for centuries to come?

*A rad is equivalent to 100 ergs of radiation energy received per gram of body weight. A millirad is therefore one-thousandth of this.

The Heat Balance of the Biosphere

The processes of generating energy have one effect in common—the energy they generate is accompanied by, and eventually becomes itself, waste heat. This is released to the biosphere, where it is superimposed upon the natural patterns of energy flow. In producing energy from fossil fuels, carbon dioxide is released, and because of this, the atmospheric concentration of carbon dioxide is rising. As we have already discussed in Chapter 2—3, the concentration of carbon dioxide appears linked to the thermal balance of the atmosphere. From the burning of fossil fuels and the processing of resources, there has also been a rise in the turbidity of the upper atmosphere. This too may be having effects upon the heat balance of Earth. Together these phenomena are aspects of a vast geophysical experiment that we are unconsciously conducting on the biosphere as a whole. What the results of this experiment will be, we can really only wait and see; at present we have inadequate evidence to make really sound predictions. And yet we can visualize trends that appear foreboding, to say the least. The results, therefore, may be anything from harmless to catastrophic. Let us consider the facts and the possibilities.

Thermal Pollution

Thermal, geothermal, and nuclear plants often produce large volumes of heated water. In some instances this is cooled by passage through large cooling towers, so that heat is lost to the atmosphere before the liquid is released to a watershed. Often, however, the heated effluent can raise the ambient temperature from anywhere between five and eight degrees centigrade. As the water temperature increases, the solubility of oxygen drops and thus the dissolved oxygen. On the other hand, many metabolic processes, including those in plants and bacteria, will speed up. Consequently, eutrophication may be accelerated.

Many fish, invertebrate, and bacterial organisms are highly susceptible to sudden changes in water temperature, particularly if they are under stress from other pollutants. Trout die at temperatures above 25°C and carp at those above 38°C. Trout eggs will not hatch at temperatures above 14·4°C. Some generating stations release heated effluent in a periodic manner; hence water temperatures in the receiving body fluctuate madly from one day to the next, giving life little opportunity to adapt. Thermal pollution of watersheds is only at a beginning stage, and its effects will be extensive should the power industry expand as predicted. So great will be the demands for cooling water that in 1980 in the United States, according to one prediction, two-thirds of the surface waters will be used thus for 50 per cent of the time. Similar parallels will occur in Canada, if allowed, although on a less intense basis because of lower population size. In one prediction of Canadian economic growth, it was estimated that the utilities sector of the economy will increase twelvefold by the year 2000, despite a predicted population increase of less than twofold.

Stoves, light bulbs, space heating, and machinery—all of these uses of energy release waste heat, mostly about urban centres. In heavily populated areas this heat is large enough to affect weather patterns. Heat islands—blocks of air with temperatures much higher than the surrounding landscape—have been observed over cities such as Toronto and Vienna. There is still inadequate data on the extent to which these heat sources may alter climatic conditions over a large area, but as urban sprawl continues, with the formation of giant strip-cities like the Boston-to-Washington megalopolis, it is quite possible that these effects could be profound.

On a global basis, the waste heat produced by humans may have important consequences, should expansion of power-production facilities continue as expected. In terms of the sunlight energy entering the atmosphere, even the 1980 production of waste heat would be insignificant—0·01 per cent. But in terms of the sunlight energy that actually reaches Earth's surface, this waste-heat energy would be large—6 per cent annually. And should the whole world consume energy on the per-capita basis of a 1970 American citizen, the waste heat would be about tenfold higher.

These estimates are very rough ones and are meant only to give a general idea of the order of magnitude of the waste heat produced by human systems. There are many factors that go unaccounted for. The oceans, for example, are very cold and will act as a great heat-sink by absorbing waste heat from the atmosphere. Although the waste heat produced by human systems is negligible beside the input of solar radiation to the atmosphere, it may have important effects upon the troposphere—the lower region of the atmosphere where weather patterns occur, for here it is of much greater magnitude.

The input of carbon dioxide to the atmosphere from the burning of fossil fuels has increased since the turn of the century, and rapidly in the last thirty years. Figure 25—1 illustrates these increases measured at locations such as the U.S. scientific station at the South Pole and Mauna Loa Observatory in Hawaii. Both of these sites are in areas of atmospheric mixing and substantial distances from the

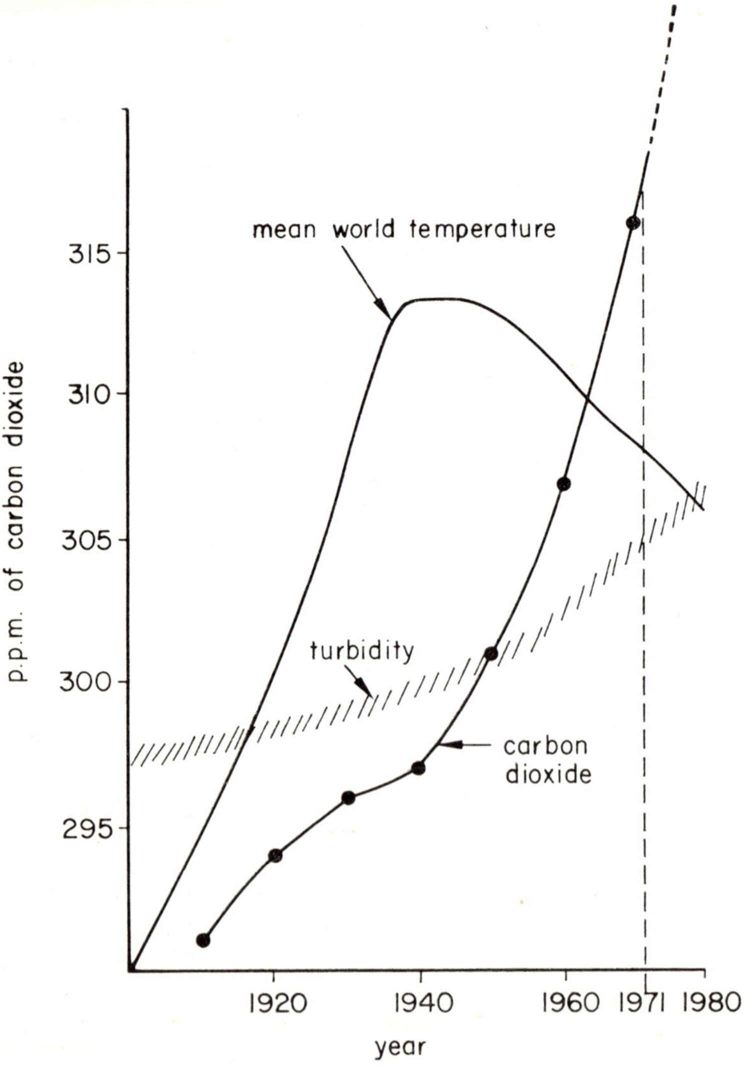

Figure 25—1 The Chilling Greenhouse

combustion of fossil fuels. The readings themselves are very accurate. In sixty years there has been a total increase of twenty-five ppm or about 10 per cent.

A gradual increase in carbon dioxide should be paralleled by a slow rise in world temperature. Between 1885 and 1940 on a world average, temperatures rose 0·5°C. Mean winter temperatures rose 0·9°C, but in the 40°-70° north latitudes they rose on the average by 1·6°C (2·8°F). From 1940 onwards world temperature has been slowly decreasing, even though more than 40 per cent of the atmospheric increase in carbon dioxide occurred in this period.

This decrease is thought to be the result of increasing turbidity in the upper atmosphere. Particulate matter from agriculture, road-building and construction, industry and jet travel is all finding its way into the stratosphere. Once up there, this matter drifts downwards only very slowly and blocks some of the sunlight energy from reaching Earth's surface; temperatures thus fall. Built of materials that heat up readily, cities often become the base for great convection currents that rise to the upper atmosphere. These currents act as great chimneys, lifting the particulate matter of urban areas upwards with them. Turbidity over urban areas has increased by amounts up to 88 per cent in the last sixty years. From the Mauna Loa Observatory, between 1957 and 1967, an increase of 30 per cent in the turbidity of the Pacific sky was observed.

In the past, fluctuations in world temperature have correlated closely with volcanic activity. Following two of the largest eruptions in recorded history—Krakatoa in 1883 and Tomboro (in Sumbawa, Indonesia) in 1816, the world experienced the lowest recorded temperatures in three hundred years. Dr Kenneth E. F. Watt has correlated these decreases with sharp rises in the price of grain on the London and Liverpool grain markets, an indication that primary productivity about the world was affected.

It appears, therefore, that the turbidity effect may have countered and overcome the increase in temperature from rising carbon dioxide levels. What the effects will be in the future is more difficult to judge; considerably more research is needed to fill in the details of the atmospheric picture. Some scientists argue that the increases and decreases in temperature are so small that they are merely due to normal fluctuation in Sun's radiation. Others argue the opposite and point to the last ice ages as a phenomenon that may have been produced by fluctuations in carbon dioxide and/or turbidity in the atmosphere. Even others feel that although the increase due to CO_2 may be valid, the decreases are not due to man-induced turbidity but due to minor volcanic activity in the last ten years.

Further, there is little known about the interrelationship between increased turbidity, increased carbon dioxide, and increased waste heat, all produced by human systems. The oceans seem to be able to absorb both carbon dioxide and waste heat; they may, therefore, be able to moderate both of these effects. It appears that, already, half of the carbon dioxide produced from the combustion of fossil fuels has been absorbed in the oceans.

In an attempt to predict what these present trends may lead to, a number of scientists have devised computer models—albeit simplified ones—of Earth's atmosphere. In these models, certain factors such as carbon dioxide content can be varied. Some of their results, although limited by our ability to make working models of the atmosphere, are interesting. For example, the projected increase in atmospheric carbon dioxide by the year 2000 will be about 25 per cent of its present concentration, giving a new level of nearly 400 ppm, a third greater than the 1900 value. Without turbidity's being considered, this could lead to an increase in average world temperature of from 0·6° to 4°C. These predictions have given rise to speculation that the polar ice-caps might begin to melt, even though such a process would take nearly 2,000 years if it were to occur completely. A more likely alternative would be the destabilization of the polar ice-packs, which could drift through the oceans, causing drastic but unknown changes of climate.

If turbidity continues to play an important role, it could lead to decreased temperatures and decreased amounts of sunlight. Growing seasons would therefore change and climatic patterns would shift. Fertile areas might decrease their productivity, and in some instances become deserts.

However, all of these predictions remain very much within the realm of speculation. If our present attempts to clean up air pollution are successful, then they will lead to a decrease in turbidity. This might unleash the greenhouse effect, which could carry the world into a crisis of increasing temperature. We must with great haste begin to attack both of these potential problems simultaneously, on both the level of basic research and practical solution.

The Availability of Energy Resources

Fossil Fuels

Fossil fuels are the remains of living matter deposited at various times during the oxygen revolution. Transformed by geological heat and pressure, some of the living molecules were broken down to simpler ones, while others were polymerized to form more complex varieties. Essentially, however, they remained uncombined with oxygen and still contained within their carbon-carbon bonds the energy fixed by millions of years of photosynthesis. Like many of the other minerals employed by man, the fossil fuels are highly localized concentrations of organic matter and are therefore discontinuous in their distribution. And again, like the richest mineral deposits, fossil-fuel deposits are located near or on the continents themselves, in sedimentary basins that were once swamps or shallow seas. By knowing the limits of these sedimentary areas on Earth, it has proven possible for geologists to estimate the total recoverable reserves of fossil fuels on this planet. The best discussion of estimates is given in the *Reources and Man* report, and some of these have been summarized in Table 25—1 below.

Table 25—1 World Reserves of Fossil Fuels

Fuel	Quantity	Time for 80% exhaustion
Crude oil	1,350-2,100 billion barrels	2020-2032
Natural gas	12,000 trillion cubic feet	2000-2020
Coal	4·3-7·6 trillion metric tons	2380

In calculating the time required for 80 per cent exhaustion of these reserves, it was assumed that new power sources would gradually replace these as well as supplying growing demands. The depletion curves in Figure 25—2 are therefore bell-shaped. Many people take issue with this sort of interpretation however, among them Dr K. E. F. Watt at the University of California. Dr Watt, a systems-analyst, argues that the depletion curve of any resource is sudden and exponential. In other words, when the end of a resource is in sight, instead of conserving it we use it all the faster. Watt points to the whaling industry as an example of this attitude. (See Chapter 2—7 for a discussion of the whale.) Watt predicts a crude-oil crisis by 1990, and before that if the supersonic transport (SST) is introduced

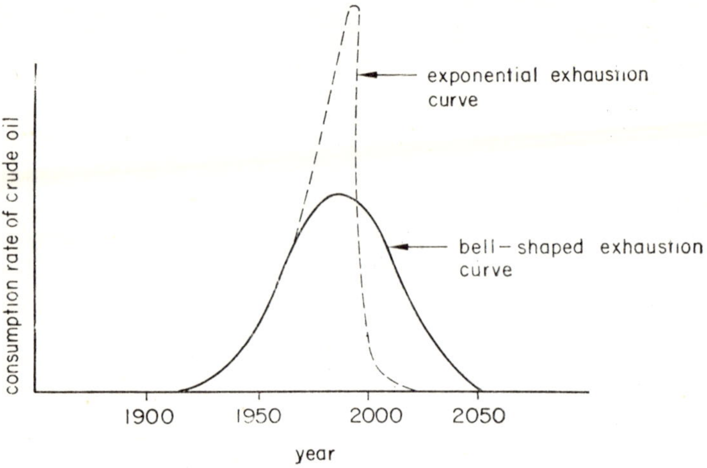

Figure 25—2 Exhaustion of World Crude Oil Reserves

because of its inordinate consumption of oil.

However, even if the estimates of Table 22—1 are fairly accurate, they give no cause for optimism. Although coal supplies appear adequate for a long time into the future, coal is the least satisfactory of fossil fuels, and in many uses cannot easily replace the liquid fuels that will be exhausted at a much earlier date.

There are other potential sources of fossil fuels that are not included in the above estimates. These include the oil-shales of the Green River region in Wyoming and the Athabaska tar-sands at Fort McMurray in Alberta. Similar deposits in other parts of the globe are known, particularly in Brazil. To extract petroleum products from these materials will require techniques that have yet to be partially, if not fully, developed. Because we have yet to experience the limitations of processing these deposits, an unfettered optimism rages today over the promise that they hold. Already evidence is accumulating that the extraction of petroleum from both the tar sands and the oil shales will be accompanied by massive ecological problems.

Geological Sources of Energy

Table 25—2 summarizes the probable total potential of hydro-electric, geothermal, and tidal power sources. (Tidal power harnesses the inrushing and outrushing tides—drawn by the moon's gravity—to drive turbines, and the first such stations are operating in France and the U.S.S.R.)

Table 25—2 Geological Sources of Electrical Power

Source	Capacity	% presently developed
Hydro-electric	2,857 thousand megawatts	5
Geothermal	60 thousand megawatts	about 2
Tidal	13 thousand megawatts	0·3

Except for hydro-electric power, the potential of these energy sources is significantly smaller than today's needs. The total geothermal potential, for example, amounts to only 20 per cent of the present electrical capacity of the United States. Hydro-electric power appears to have the largest potential, but even if we develop half of this, it would only double today's production. However, some of the best sites for hydro power are of great scenic and ecological value, and it is unlikely that we will sacrifice all of these for one to two hundred years of power. The largest potential for hydro power lies in Africa and South America, yet tropical ecology presents formidable problems to development.

Solar Energy

At first glance solar energy appears to be the logical and almost unlimited source of energy for human demands. Perhaps if we were to trap Sun's energy, we could increase our present generating capacity ten times over. But to harness this energy is far from easy, because it is dispersed over very large areas. To collect enough energy to operate a 1,000 megawatt electric plant, for example (enough to provide for a city of 1·5 million people and about the average size of many generating stations today) would require a surface area of sixteen square miles! Additionally, to construct sunlight receptors large quantities of rare metals such as cadmium and mercury are needed and, along with the high cost of equipment and land for a single plant alone, this seriously limits the practicality of such a venture. Unless some radical new technology is developed in the near future, solar energy will not provide an answer to even our present demands. With the good design of homes, however, sunlight can be employed for space-heating, a purpose that consumes 10 per cent of today's energy budget.

Nuclear Power

Even though the nuclear-power program has barely got underway, reserves of fissionable uranium hold surprisingly little promise, particularly in light of the planned expansion of the reactor program. In an ore body of uranium, only about 1 in 141 uranium atoms is a fissionable 235 isotope. (The other atoms are predominantly infertile uranium-238.) Today's estimates of the total extent of world uranium reserves are somewhat limited by lack of knowledge. However, the projected United States' demand of 675,000 short tons of uranium-235 oxide by the end of the century is more than the total estimated U.S. reserves, and more than a third of total world reserves. By 1980 alone, 275,000 tons will be required, of which only 210,000 will be available. These are reserves that are extractable at ten dollars a pound, and although reserves at a higher cost of production do exist, use of these could create problems in the power industry. Yet even if these reserves were tapped, it is questionable whether they would last for very long under the sky-rocketing demands of the power industry.

The nuclear-power industries have a proposed solution to the problem of growth in the industries' threat to deplete uranium reserves. Just as nuclear fission, they say, will carry humanity out of a crisis of fossil-fuel shortages, so will the breeder reactor and fusion power lift us out of a crisis in the availability of nuclear fuels.

The Breeder Reactor and the Fusion Reactor

Although naturally-occurring uranium-238 is not itself fissionable, the use of small amounts of uranium-235 makes it possible to convert it to plutonium-239, which like uranium-235 is fissionable. Similarly, non-fissionable thorium-232 can be converted to fissionable uranium-233. Both of these starting materials (uranium-238 and thorium-232) occur naturally to a much larger extent than uranium-235. These conversions may be carried out in a nuclear reactor.

If, in the process, only the uranium-235 that is added in the first place fissions, then the reactor is merely a *burner*. If, however, conditions are adjusted so that some of the uranium-238 or thorium-232 is simultaneously converted to fissionable material, then the reactor is a *converter*. And if it converts more of this non-fissionable material than there was uranium-235 in the first place, then it is a *breeder* reactor. It is upon the breeder reactor that hopes for the next thirty years are being placed. Even though converter-type reactors will perhaps double the lifetimes of our fissionable uranium reserves, by allowing the use of some thorium-232 and uranium-238, unless we are able to breed a net amount of fissionable material, then we shall eventually burn up all of our uranium reserves. Attempts to develop safe and economical breeders are underway, but they are not yet a reality, and even if they should be introduced, as it is hoped by 1985 on a commercial scale in the United States, a period of ten to twenty years would be necessary before the initial fuel inventory could be doubled. Breeder technology is potentially a far more dangerous method of generating power than the conventional fission reactors. But in any case, breeders will only extend the lifetime of fossil fuels.

In the process of fusion the nuclei of light atoms are fused together (rather than heavy elements split apart as in fission) and energy is released. Although the reaction has been carried out in the form of the hydrogen bomb, it has so far been only in an uncontrollable manner, and research today is attempting to control the process and then to harness it. In particular it is hoped that the fusion between deuterium (hydrogen-2 atoms and deuterium atoms, and that between deuterium and tritium (hydrogen-3) can be controlled. Deuterium occurs in sea-water as 1 in every 6,500 hydrogen atoms and this is economical to extract. The oceans therefore hold vast potential for the development of fusion power.

Yet there are still many barriers to the ultimate realization of a safe and controlled fusion reactor. Some physicists feel that intrinsic instabilities within the process itself preclude its eventual use as a power source. There is still no guarantee that the process is free from environmental side-effects, and radioactive wastes like tritium will be produced and must be handled properly. Waste heat will also be produced.

Patterns and Politics of Energy

Energy demands and sources, like those of material goods, underline the political policies of the developed nations. Again, a large portion of the energy used annually is consumed by a small proportion of the world's populace. And again, these developed nations, enamoured with economic growth, are hoping to double their energy consumption in about twenty years. In an earlier part of this chapter, we viewed some predicted lifetimes for some of the fossil-fuel reserves on this planet. It is possible, assuming that consumption of oil continues to rise in the Western world and is not substantially replaced with nuclear power, to make some calculations of the potential lifetimes of world oil reserves under a number of conditions. Table 25—3 illustrates these:

Table 25—3 Lifetimes of Oil Reserves

Conditions	Time in years from 1971 to 80% exhaustion
1. Predicted by M. K. Hubbert in *Resources and Man*	48-61 years
2. World at predicted 1975 consumption rate (minus U.S.A.)	148 years
3. World at predicted 1975 U.S.A. rate of consumption	12 years

The first estimate is based upon the assumption that oil use will ease off as new sources become available. Let us assume, however, that consumption does not decline, and the 1975 world consumes oil at the projected 1975 rate, 2·4 barrels per person annually, without the U.S.A. The 80 per cent exhaustion point would be reached in about a century and a half. In comparison, the 1975 per-capita U.S. rate is projected at thirty barrels per person per year. Should the 1975 world population consume oil at this rate, then oil reserves would last little more than a decade. Here then lies the great disparity in energy production. If our oil reserves at the present moment do not appear in a very tenuous position, it is simply because the oil pie is unevenly divided. We would face immediate problems if attempts were made to correct this situation. Although the predicament outlined here was applied to oil reserves, it could be applied also to gas, coal, and uranium reserves.

Energy and the Technological Fix

The human race is approaching the limits set by the availability and the use of energy within a closed biosphere. Certain energy resources such as the fossil fuels and uranium are in serious danger of over-exploitation should present and projected trends continue. The effects of energy generation are being felt about the biosphere, through hydro projects, coal mining, oil transport, and the release of carbon dioxide, radioactivity, and heat. These effects may even become catastrophic should present trends continue.

Despite the potential of the fusion and the breeder reactor, we must not become over-optimistic. Simply because we have had some success with nuclear programs to this stage does not automatically mean that human technology will overcome the problems associated with these new methods. Government and power companies, however, claim the opposite. The breeder reactor, and eventually the fusion reactor, they say, will be the answer to all our problems. Not only will they provide us with very large amounts of energy for very long periods of time, but they will be virtually pollution-free. But these are views of individuals undaunted by the failures encountered in the atomic-energy program from the fifties to the seventies. Plants have not opened on time, and side-effects that were not initially expected have been noted.

Can we anticipate a completely pollution-free technology? There are many scientists who feel the opposite of the power people, who feel that processes like the fusion reactor will only intensify the release of radioactive wastes into the environment; and of course at increased levels, for Western society is not attempting merely to meet the disappearance of fossil fuels or to replace conventional hydro or thermal generating facilities with nuclear ones, but is striving to meet the increases demanded by economic growth. The energy supply-and-demand situation of North America alone has reached a very tense period. Since conventional methods of energy production are not satisfactory in meeting increased demands, the plans for future development are being formulated in terms of projected technologies *that do not exist*. There is every possibility that we may develop these processes, yet the converse is equally true, for it is impossible to assess a project adequately until we know how to carry out that process. So great is the faith in technology today, that we have no qualms in planning the human future and justifying our present actions on the basis of processes that have yet to be partially, if not completely, developed. This must be the folly of all time.

For the future of the whole of human society we are going to require safe and guaranteed sources of relatively clean power. Our dealings with energy will always produce waste heat, and unless radically new methods arise, the generation of energy will always have some effect upon the biosphere. Energy production must, therefore, be limited to the long-term availability of energy resources and to the capacity of the biosphere to handle the stresses that it creates, and it must only be used for sound development. Technology may help us to do all of this, but in order for this to happen we must have full mastery over this technology and the power that it produces.

CHAPTER 2—6

The Phosphorus Sump

Phosphorus is an element vital to life. Combined with four oxygen atoms to give the phosphate ion (PO_4^{3-}), this is the form in which phosphorus is commonly found in the biosphere. Phosphates are either combined with atoms like sodium (Na_3PO_4), calcium ($Ca_3(PO_4)_2$), or hydrogen (NaH_2PO_4) in the non-living parts of the biosphere, or in a wide variety of forms in living matter. Calcium phosphate, for example, is a major constituent of teeth and bones. Fats called *phospholipids* consist of a phosphate group attached to a lipid moiety and are essential to the structure of cell membranes. Sugars such as glucose and fructose are used by cells in a form combined to a phosphate group. In the nuclei of cells, phosphate linkages are an important feature of the molecules that store the genetic messages. And paramount in biochemical importance are the triphosphate linkages used by all cells in the processes of energy use and storage. These energy-rich bonds are formed during the breakdown of glucose and transport the chemical energy to the actual site of demand, whether it be muscle contraction, biosynthesis of complex molecules, or nervous sensation.

The Dynamics of Phosphorus

Phosphorus in the biosphere is unusual. Unlike the elements of the other cycles that we have looked at earlier, phosphorus remains fundamentally in one chemical form, that of the phosphate group. As Figure 26—1 shows, phosphate erodes from rock and is carried into the soils and eventually into rivers and streams. In the soils it is absorbed by fungi, bacteria, and plants as phosphate; it is then biochemically utilized as phosphate. In all of its biological lifetime in the food chains, although it is combined with many different chemical groups, phosphorus remains in the phosphate form. And finally, through decomposition, it is released as phosphate again. Here it is taken up by living organisms or washed into rivers and lakes. During its passage through watersheds it may be reabsorbed and released again by the aquatic food chains. It may be precipitated out of the lake as insoluble ferric phosphate and eventually be released once more as soluble ferrous phosphate. Finally it reaches the oceans, and gradually, as dead life seeps into the ocean muds and is carried away without decomposition, it is carried away also, particularly in the insoluble remains of shellfish, skeletons, and other marine organisms.

There is, thus, a slow but net flow of phosphorus from the land masses where it is worn away by erosion, to the deep sea-depths where it is buried in the sediments and taken out of the biosphere. With the carbon cycle there are gaseous components such as carbon dioxide and to a certain extent methane. With the nitrogen cycle there are gaseous components such as nitrogen, nitrogen oxides, and ammonia. With the sulfur cycle there are gaseous sulfur dioxide and traces of hydrogen sulfide. However, with the phosphate cycle there is really only the phosphate form, which is to all intents and purposes a water soluble-solid. Occasionally, anaerobic decomposition of living material will release phosphine (PH_3), but this gaseous component is highly unstable in the presence of oxygen and ignites to produce phosphorus oxides, which again end up as phosphate. (Sometimes one sees flickering lights over a marsh at night; these will-o'-the-wisps are spon-

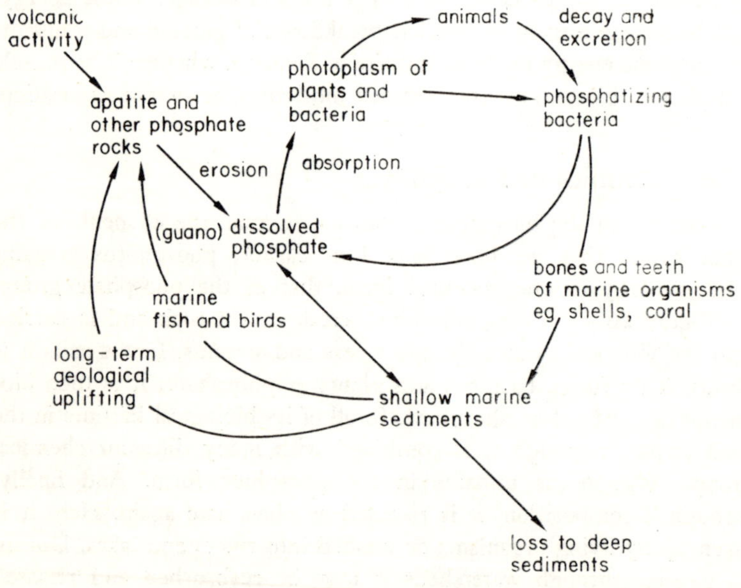

Figure 26—1 The Phosphate Cycle

taneously igniting mixtures of methane and phosphine.) Because there is essentially no gaseous phase to the phosphorus cycle, it has little buoyancy to speak of. Unable to return to the atmosphere through volcanic discharge and through decomposition, it is forced downwards to the deep ocean sediments. The phosphorus cycle then, appears to be unbalanced.

Over the long term, however, there is some reversal of this cycle. Large amounts of the phosphate lost to ocean sediments is lost in shallow coastal waters. Here it remains fixed in the form of bone and skeletal remains and, in warm tropical seas, in the form of coral reefs. During geological time many of these have been slowly uplifted and are now parts of the land masses. The fossilized remains of these organisms, compacted beneath layers of sediment, have become phosphate-rich rocks. It is these rocks that provide much of the phosphate entering the biosphere through erosion.

To retard this downward flow of nutrients, life has apparently devised many mechanisms working at all levels of the biosphere. In the soils, phosphate is bound very strongly to clay particles and even more strongly to the particles of clay-humus that are formed during the process of soil building. Up to 85 per cent of the phosphates in soils may be held in this way. As a result, in temperate areas stable soils may contain about 650 ppm of phosphate, while the soil solutions (the ground-water within the soil) contain between 0·001 and 30 ppm of phosphate. The phosphate that is bound to these clay-humus particles is resistant to bacterial and fungal action. So too is the insoluble phosphate contained in animal bone, or in minerals such as apatite. From all these reserves soil phosphates are only very slowly freed, so that their release can be accompanied by a slow and steady uptake of phosphate into the green plants, with minimal losses from the soils.

Green plants are continually transpiring or losing water through the pores of their leaves. Water is replenished by absorption through the roots. One could picture a plant as a vessel through which water flows from the soils to the atmosphere, at slow rates. As water is absorbed into plants, dissolved nutrients such as phosphate, nitrate, and calcium are carried along with it. These do not leave the plant through transpiration; they remain in the vegetation. Thus, transpiration is a mechanism by which green plants concentrate nutrients from the soil and ecologically fix them into a form that is not easily washed away. Phosphate may be concentrated in green plants to levels as much as 1,500 times that of the soils. Phosphorus appears to be concentrated more than any other nutrient.

As these green plants are consumed by herbivores, and these by secondary and tertiary consumers—especially if they are vertebrates—the phosphorus is concentrated even more. In these organisms it is fixed to a very high level in bones; bones contain 100 times the concentration of phosphates in the surrounding tissues. Following decomposition these bones form long-term repositories of phosphate in the soils. Otherwise, the phosphate released by decomposition is reabsorbed rapidly by living organisms or by clay-humus particles. In tropical ecosystems where heavy rainfall will quickly leach out nutrients, phosphates, like nitrates, sulfates, and calcium, are snatched up by extremely rapid mechanisms. Most of the phosphorus reserves are here maintained within the bio-mass itself.

In many temperate fresh-water lakes, particularly nutrient-rich or eutrophic ones, there appears to be some sort of phosphorus regulation. The lifetime of free inorganic phosphate is often extremely short—it is snatched up quickly by phytoplankton and other organisms. Nearly all of the phosphate is associated with particulate and organic matter that is either suspended in the water or constitutes, with mineral particles, the sediments. And as in soils, in the sediments phosphate will cling to the clay-humus particles. In some lakes the biological usage of phosphate is linked cyclically to the seasons. In winter, when the lake is frozen over, the phosphorus originally held within the phytoplankton community remains tied to organic material which settles in the bottom muds. While the deposition of this matter is occurring, in late summer and early autumn, the waters of the lake are turbulent and oxygen-rich. The iron dissolved in them is mostly in the ferric (Fe^{3+}) form. At the mud surfaces, the ferric iron seems to react with the phosphate to produce insoluble ferric phosphate which precipitates out. However, during the winter there is little wind or water motion below the ice, and the decomposer bacteria begin to deplete the oxygen of the bottom layers. Anaerobic conditions start to predominate, and the sulfate bacteria reduce sulfate to sulfide, while the iron bacteria reduce ferric to ferrous (Fe^{2+}) iron. Ferrous phosphate is soluble, and so, it seems, phosphate is released again. This process appears to be maximized in spring, so that large amounts are released to supply the new boom of plankton. If it were not for this annual deposition of phosphate from the bio-mass in the form of insoluble sediments, then the system would annually suffer large losses.

In the oceans, further mechanisms are apparently at work. Marine plants, both seaweeds and phytoplankton, concentrate phosphorus to a very large degree above that of the surrounding environments; concentrations from 20,000 to 15,000 times greater have been observed. At the end of the marine food chains are invertebrate and vertebrate skeletons that act as phosphate storage banks. These deposits are more susceptible to burial and to loss in deep sediments than their terrestrial counterparts, but in coastal areas, estuaries, and shallow seas, they help for a short while to hold back the phosphate flow. Invertebrate skeletons from sponges, barnacles, corals, sea fans, starfish, sea urchins, oysters, and other molluscs all contain a fairly high percentage of phosphorus (about 2,500 ppm), although their basic constituency is limestone, not the apatite of vertebrate bones. Although phosphorus is removed at a slower rate from sea-water than it is from fresh water, it is still removed quickly, and as we noted in the earlier discussions of the biological control of the environment (Chapter 1—3), sea-water phosphate probably plays a major role.

Finally, as shown in Figure 16—1, there are pathways by which phosphorus is biologically returned to the land masses. Salmon return deep into the land masses to spawn and die, and thus their skeletal remains are returned. Vast numbers of marine birds make their catch at sea and return to the land masses to feed their young. The remnants of this process are phosphate-rich bones and guano, deposited along cliffs and coastlines. Many oceanic islands are heavily laden with these deposits, particularly off the coast of Peru.

We cannot say for certain that life is attempting consciously to slow the loss of phosphorus to the ocean sediments. However, had living systems not developed methods to retain this essential but scarce element, then it is likely that many ecosystems would have starved from phosphate shortage. In other words, there may have been a distinct selective advantage to ecosystems that could conserve this element. It is difficult to tell though, with any of the biogeochemical processes of life such as phosphate cycling, whether they were a chance (and very fortunate) consequence of the ways in which life operates, or whether they were a result of evolutionary selection of the sort that occurred at the level of species.

Double-Headed Misbalance

Human activity interferes with many of the mechanisms that slow the loss of phosphorus to the sea. For example, phosphate-rich rocks, which, biologically, would have undergone slow erosion and fertilized many centuries of ecosystem, are being mined. One of the demands for these rocks is agricultural fertilizer; another is detergents. The need for fertilizers is a product of agricultural technology. Disrupting the root systems of grasslands and forests in order to plant crops has caused very drastic changes in soil ecosystems. Greater leaching and loss of nutrients occurs. Fertilizers are therefore spread upon the fields to replenish the eroded phosphates. Many of the retention mechanisms of the system have been disrupted and the fertilizer phosphate is often washed away quickly into lakes and streams, particularly if applied during periods when little growth is occurring. Some 'slow-release' fertilizers consist of bone meal and/or apatite. These fertilizers mimic biological-retention mechanisms.

After the crop ripens, the fields are harvested. Little of the plant matter remains, most being used for food or other purposes such as insulation, animal bedding, and clothing. Large numbers of animals such as cattle and pigs are often kept in enclosures or compounds of incredibly small size. People are kept in similar enclosures, termed cities. The volume of metabolic wastes from these colonies is very large and localized. It has been estimated, for example, that the livestock of the United States produces a volume of sewage equivalent to two billion humans! Rather than being returned to the land ecosystems from whence it was derived, these wastes are usually released into watersheds. Occasionally treated to remove pathogenic bacteria or organic matter, they are rarely treated to remove phosphates. Thus the phosphates from these metabolic wastes flow almost directly into the watersheds to join the phosphates from erosion caused by agriculture and forestry, and from the use of phosphate-rich fertilizers.

At this point there is another input of phosphates, derived again from phosphate rock. This rock is mined and chemically converted to polyphosphates, which have proven to be a rather successful component in modern laundry detergents. In dense urban areas of the developed world large amounts of these detergents are being used, and they are released with human sewage.

In all of these ways, human activity is accelerating the one-way flow of the phosphate cycle. Although very little work has been done on the phosphate budget of Earth, some estimates have and can be made of the degree to which human affairs are altering the situation.

In 1948, when the annual production of phosphate rock fertilizers was about ten million tons, G. Evelyn Hutchinson, the eminent ecologist, estimated that humans were accelerating the natural flow of phosphorus by 5-10 per cent. Today, when the production and usage of phosphate fertilizers is much higher, so must be this effect.

According to United Nations figures, the world consumption of phosphate rock fertilizers in 1968 amounted to 63 million metric tons, or 7·6 million metric tons of elemental phosphorus. At the same time, the natural loss of phosphorus in the biosphere from the land masses to the oceans is about 14 million metric tons per year. In other words human activities have raised this by 50 per cent, 5-10 times the rate twenty years ago!

Total world reserves of phosphate rock (including lower grade and presently not mined rock) total about 3,140 million tons of phosphorus. At present rates of consumption these will last some 400 years. Phosphate rock however is used far more by the developed nations than it is by the developing nations. To bring the whole world to a level of per capita equality with the developed sector would require more than twice as much phosphate per year than we now consume, and would shorten the lifetime of world reserves to only 160 years. Furthermore, if we assume that population growth will continue, then phosphate demands will also rise, and the lifetime of the reserves become shorter still.

The results of such an analysis are frightening. To feed the growing population of today requires increasingly more intensive agricultural technology with a much greater reliance on fertilizers. With current growth rates of population, and of phosphate-rock consumption, we may find our known reserves rapidly dwindling in another half a century, while we are stranded with a population of 11 billion people to feed!

Phosphorus will, and always will continue to be an element essential to life. In the dynamics of human interaction with the biogeochemical cycling of phosphorus, we are threatening the future of not only ourselves but of the whole of life. It is essential that we understand as swiftly as possible the dynamics of phosporus and life and the limitations which the mechanisms of life attempt to overcome or to prevent. We must find ways of working in parallel with those mechanisms; we must find ways of holding back phosphorus on the land masses, of fully utilizing all of the rock that we mine, instead of simply extracting the higher grade material and discarding the rest. We must return the phosphorus-rich sewage from human settlements and livestock enclosures to the land, and we must shift our agricultural technologies to methods that maintain the nutrient content of the soil without the need of heavy applications of fertilizers; failure in all of these may mean the slow diminution of life processes on Earth as phosphate litters the ocean depths as an insoluble and irretrievable sediment.

CHAPTER 2—7

The Oceans

To the human mind, the oceans have long been held as boundless. To one at sea, the waters appear infinite and the horizons stretch flat as far as the eye can see. We have come to see the oceans as cornucopias, containing everything that is limited on land—food, minerals, cheap water, oil and gas, power, and even as receptacles that in their infinite size can handle our wastes.

We have seen already, however, that the oceans are not likely to provide us with very much more food than we are consuming now, although by a little ingenuity and with more widespread but carefully regulated exploitation we may be able to increase our present yield by about 60 per cent. Similarly, we have seen that the oceans will hardly be an infinite source of minerals and energy. We may be able to exploit deposits on the continental shelves such as oil; perhaps we shall be able to extract several elements from sea water economically, but nothing to the degree that has been envisioned. And the oceans are hardly infinite in their capacity to bear the brunt of human industrial wastes and human exploitation. Danger signs are beginning to appear.

Pesticides, Mercury, and Oil

Of the persistent pesticides entering marine ecosystems, DDT is the best studied. Several inputs, of many possible, have been examined. Dust in the north-east trade winds, for example, carries about 600 kilograms of pesticides annually into the equatorial Atlantic. The San Joachim River of California carries about 1,900 kilograms annually into San Francisco Bay, and the Mississippi brings a million annually into the Gulf of Mexico.

Beyond the coastal areas, winds, tides, currents, and rainfall all carry DDT outwards towards the middle of the oceans. Evidence of this are the high residues of DDT found in the Bermuda petrel. Apart from nesting on the far eastward cays of Bermuda, this is a bird that lives its life in the mid Atlantic. Its eggs are now thin-shelled and contain high concentrations of DDT. In many species of ocean-going fish, notably tuna, mackerel, and hake, DDT levels of between 0·2 and 2·0 ppm have been found.

Perhaps one of the most bizarre effects of DDT yet noticed is that on phytoplankton. Dr C. F. Wurster, Jr (State University of New York) has demonstrated that DDT in only a few parts per billion can drastically reduce the rates of photosynthesis of five phytoplankton species that are commonly found in Atlantic communities. Although nothing like these concentrations of DDT have been found in sea water, levels of DDT as high as this have been found within phytoplankton themselves, through accumulation. We cannot as yet tell whether this is the same thing, but it could warn of a massive change in the ecology of the oceans, as more-resistant phytoplankton communities become predominant and less-resistant ones shrink.

Many other exotic chemicals are following paths similar to those of DDT. These include the pesticides aldrin and dieldrin and PCB's, which have been found in high levels in birds and fish off European coasts. We must look with great urgency towards tropical marine ecosystems, for these border on lands where large amounts of DDT are used for the control of malaria.

Mercury too is showing up in the food chains of marine ecosystems. It is debatable whether all of this mercury arose from human activity; some of it may simply be concentration of background mercury in the sea water. The major human sources of mercury are pulp-and-paper installations and fertilizer plants. One of the first cases of marine mercury pollution was that of Minamata in Japan where many people died from eating fish and shellfish contaminated with mercury. Already the Baltic has been polluted with mercury from Swedish pulp-and-paper mills, and James Bay from mills in Northern Ontario and Manitoba. Extensive pollution probably exists in other parts of the world where pulp-and-paper industries are located, such as the Fraser River and Puget Sound, the Sea of Japan, the east coast of the United States, and the coastal waters of Chile. Tuna and swordfish in the Atlantic have been found with mercury levels above the acceptable standards of the World Health Organization (0·05 ppm), and some even exceed the U.S. standard of 0·5 ppm. Although it has not yet been reported, West Coast salmon stocks are likely also contaminated, since they spawn in the effluent of some of the world's largest pulp-and-paper concerns. Although it is certain that much of the mercury contamination of fish in coastal and inland waters is due to industrial wastes, the degree to which ocean-going fish are contaminated by mercury from geological sources as opposed to pollution from human activity is not quite certain.

The current search for oil deposits has led to exploratory drilling and extraction procedures on nearly all of the world's continental shelves. Current oil demands have led to the extensive transport of oil by sea, as oil is shipped from Venezuela and the Middle East to North America, Japan, and Europe. All of this activity has led to a contamination of coastlines from a large number of conditions—runaway wells, inadequate safety precautions, overloading of oil cargo vessels, poor conditions of vessels, and unavoidable chance accidents.

Perhaps the best-known cases of pollution from oil-drilling have occurred off the continental United States. The 1969 blow-out off the Louisiana coast is such an example. One of the drilling rigs, owned by Chevron, caught fire, and eight wells burst loose. Forty days later two wells were still out of action and releasing 30,000 gallons of crude oil daily. An oil slick covered an area of seventy square miles and threatened to damage the oyster and shrimp beds upon which the Louisiana fisheries depend.

Another disaster was the blow-out in the waters of Santa Barbara, California, in 1969. Strains in the rock in which drilling was conducted appear to have caused cracks through which the oil now continually seeps. Although these two disasters are the most publicized, there are numerous reports of minor spills about the world. A U.S. coast-guard officer on the Gulf Coast, for example, reports the sighting of new oil slicks daily and about twice a week these are thirty-two square miles or greater in size.

Spills from the shipping of oil are a common occurrence. In 1966 some 700 million gallons of oil were transported by sea and probably a million of these were spilt. Somewhere between one million and one hundred million tons of oil are added to the oceans annually. In U.S. waters alone there were, in 1966, at least 2,000 oil spills, and the number has now increased to something like 75,000 annually. Part of the spillage results from the transfer of oil to tankers and coastal facilities. Larger amounts result from tanker and barge accidents and tanker bilge washings (which are illegal). The most publicized collisions and groundings have occurred off Europe and North America. They include the *Torrey Canyon* off the south shore of Britain, the *Arrow* off Chedebucto Bay, Nova Scotia, and tankers in

San Francisco Bay. In the Chedebucto case, the *Arrow,* a tanker owned by Aristotle Onassis and of Liberian registry—which means that it does not have to conform to the strict requirements of British registry—ran on the rocks in the fog. On closer examination, parts of its hull were found in a weak and rotten condition.

The effects of oil are widespread, the most noticeable being the fouling of miles of coastline. The feathers of birds become clogged with oil so that they lose buoyancy and are unable to fly. Following the weakening effects of the oil they are highly susceptible to diseases such as pneumonia. Only a small percentage of the birds that were cleaned following the *Torrey Canyon* and the Santa Barbara spills survived more than a month. In 1957, the *Tampico Mara* grounded in a cove off Baja, California, and the oil spill destroyed nearly all the life of the cove. Even ten years later conditions were not back to normal.

As the oil spreads itself over the surface of the waters, phytoplankton communities are smothered. Gradually the lighter fractions evaporate and the heavier ones are worked upon by fungi and bacteria. Some of these heavier fractions become adsorbed to particulate matter and sink, particularly in still waters. And decomposition of these on the sea-floor leads to anaerobic conditions, so that methane bubbles to the surface.

Some of the components of oil may have important biochemical effects. For instance, some of the chemicals may be similar to those used by marine invertebrates as behaviour signals during hunting, escape from predators, homing, and mating. Responses to these molecules require only minute concentrations, and should these molecules or biochemically similar ones be present in oil, they may drive many creatures to berserk deaths. Other components appear to resist bacterial decay; these, like benzopyrenes, are highly soluble in living tissue and may concentrate biologically. Some of these are also potent carcinogens. The extent to which both of these effects are occurring desperately needs more research. Indeed biological observations in several instances following oil spills suggest that these more subtle effects are very possible.

Overfishing

Certain major fisheries have begun to decline in the last twenty years and to date there has been no major recovery. These include the East

Asian sardine, the Californian sardine, the North-west Pacific salmon, the Atlanto-Scandian salmon, and the Barents Sea cod. Other species, such as the Newfoundland cod, the North Sea herring, the menhaden, the British Columbia herring, the Bering Sea flatfish, the yellowfin tuna, and the North Sea plaice are also showing signs of overfishing. Fisheries are not responding to declining fish stocks by pushing for protective legislation and reduced catches that will allow the stocks to recover. Instead they are employing bigger and better technology in an effort to exploit the last remaining schools as efficiently as possible. Canada has recently moved to restrict fishing of the endangered Atlantic salmon, but Denmark has refused to co-operate on the high seas.

The most frustrating story of overfishing is that of the whale. After exhausting the giant blue whale stocks, the whalers turned to the smaller fin whale, then to the Sei whale, and finally to the sperm whale. Between 1933 and 1966, although the whale catch doubled the yield of oil decreased by 40 per cent. Better and bigger whaling vessels, more advanced sonar devices—none of these were of any avail, for the shift to the smaller species of whale meant lower yields of oil. This decrease in productivity, however, was not without ample warning. Following the Second World War the nations involved in whaling joined together to form the International Whaling Commission (I.W.C.), a body to regulate whaling and to set limits on catches. In 1960 they appointed a team of biologists to investigate the situation. The committee reported in 1963 that the blue and humpback whales were near extinction and should be given protection. Furthermore, they estimated that if unrestricted whaling was permitted in 1964, the catch would not surpass 8,500 blues and 1,400 fins. The whaling industry set a quota of 10,000 blues. Their quota was never met, for their catch was virtually that predicted by the biologists. And the total catch of fin whales in that year was estimated at 35 per cent of the total number alive, a catch three times greater than that allowable on a sustained-yield basis. In 1964, the biologists recommended a quota of 4,000—the I.W.C. accepted 8,000. Barely 7,000 units were caught. In 1965 a quota of 4,500 was set—again above the recommendation of the biologists—and barely 4,000 were caught. From 1960 onwards, sperm whales constituted more and more of the catch as the yield of bigger species dropped. Finally, at the end of the decade, the catch of sperm whales crashed precipitously. During the sixties, declining catches pushed both Norway and the Netherlands out of the whaling business leaving only Japan and Russia as major contenders. Still today, no concerted international action has been taken to protect the whale. The whaling nations have refused blankly to restrict their activities. And today Japan, desperate for protein, is exploiting porpoise stocks.

The Threatened Reefs

The coral reef ecosystem is characteristic of warm and shallow tropical seas. Reef ecosystems have some of the highest known productivities of any global ecosystems. And they are the result of coral organisms. Biologically, coral organisms include some of the oldest invertebrates around, and fossilized reefs as old as 450 million years have been discovered. Coral animals are tiny unprotected creatures (somewhat like jellyfish) called 'polyps'. Each polyp secretes about itself a protective limestone—silicate shell. Hundreds of polyps aggregate to form colonies—a piece of coral—and thousands of these pieces aggregate together to form a community—the coral reef.

Coral reefs form in a process akin to that of plant-community succession. Starting in the shallowest areas of warm seas, colonies slowly build up. As the years go by, individual organisms die, but their shells remain, stabilizing the sediment about them. Their descendants build on top of them, and their descendants upon them. Over hundreds of years the reef rises until it reaches the surface of the ocean, where an equilibrium is attained. Parts of the surface may become covered with sand and other debris. Eventually plant seeds and small animals arrive, brought by the wind, waves, and rotting logs. Sea-going birds also arrive. Thus communities develop upon the reefs. Many of the coral islands and atolls of the Pacific arose in this way. Other reefs formed in the shallow waters about volcanic islands.

If it were not for the continual building action of the new generations of coral polyps, the sea would quickly erode away the reef. Many of the volcanic islands in the Pacific are surrounded by reefs that built upon the underwater shoulders of the mountains. These reefs buffer the action of the seas, creating quiet lagoons between the coral and the shore. Consequently, as wind and rain slowly erode the rock of these volcanic isles, and rainfall carries this debris seawards, it settles there into flat fertile plains. If it were not for the coral organisms, the coastline would be constantly changing and the eroded sediments would be washed away.

The coral is also a potent ecological force. The many crannies in the coral reef provide homes and protection for a diverse range of marine animals, from sea urchins and starfish to molluscs such as the oyster and the octopus, crustaceans such as the barnacle and the crab, and many fish species. The sheltered nature of the reef provides peaceful waters away from ocean currents and a productive community with ample food organisms. It is no wonder then that the reefs are areas for the spawning and raising of the young of many ocean-going fish. The reef ecosystem is a priceless link between the deep waters of mid-ocean with their low productivity and the shallow and rich waters about tropical islands.

The coral polyp has few known predators. One of them, and until recently very rare, is a peculiar starfish called 'crown of thorns' (*Acanthaster planci*). Unlike most starfish this one is venomous, being covered with inch-long poisonous spines. It feeds by wrapping itself about coral and remaining there until it has digested and absorbed the polyps—usually about six to eight hours. Suction cups on its underside allow it to climb the steepest of reef faces. Hordes of these starfish can move in great strips at about ten miles per year, leaving the coral white and lifeless behind them. Quickly these skeletons become overgrown with algae and it is no longer possible for the polyps to reinvade them.

In various locations across the Pacific during the last twenty years, population explosions of these starfish have occurred. There are two apparent reasons for this, both of them being human factors. First, one of the only known predators of this organism is a mollusc called the 'triton', whose shell has long been the object of collectors. The first outbreaks of *Acanthaster* were noted near Green Island on the Great Barrier Reef where, between 1949 and 1959, over 100,000 specimens of this rare shell were removed. Secondly, one of the predators of the starfish is the coral polyp itself, which feeds on young starfish larvae. As long as the reef is all coral, it is unlikely that many of the larvae will survive to the adult stage. But the dredging of channels and harbour, either by mechanical means or by blasting, has destroyed large strips of coral and provided areas of refuge for *Acanthaster* to multiply.

In the past decade these starfish have virtually destroyed 300 miles of Australia's thousand mile Great Barrier Reef. Further, they have appeared in the Tahiti, Fiji, Marshall, and Marian islands (see Figure 27—1) and in the waters of Indonesia, New Guinea, and New

Caledonia. On Guam they have destroyed a quarter of the reef ecosystem. The building action of the polyps is essential to the productivity and the stability of the reefs. Without it they will erode away under the action of the waves; eventually many of the islands may disappear under the seas.

Efforts are being made to control the starfish. Divers armed with formaldehyde-loaded hypodermic guns have halted their advance on Guam. Scientists are attempting to breed the triton in laboratories as well as searching for other predators of the starfish. But they are racing against time, because the reef systems of the Pacific are so extensive, infestations are so widespread, and really adequate means of dealing with this problem are still not available.

The Oceans and the Future

Figure 27—1 shows the areas of primary productivity in the Pacific Ocean. The majority of ocean waters are ecological deserts—productivity is in the order of fifty grams of carbon fixed per square metre per year. In coastal waters this can rise to 100, and in areas of upwelling of currents this can triple. About tropical reefs productivity can be high—one reef measured in the Marshall Islands reached 4,900 grams. These reefs act as focal points in a band of productivity stretching from Tahiti to Malaysia.

The mainstays of this network of very productive and desert ecosystems appear to be receiving the brunt of human treatment. Consider, for example, the productive strip of waters up the west coast of the U.S.A. and Canada to Alaska, waters that receive pulp-and-paper wastes, mining and processing wastes, oil drilling, municipal sewage, and intensive fishing. The Sea of Japan is in a similar position.

The total picture of the Pacific is far from encouraging, and yet it is only representative of what is happening to the oceans on a global scale. Productive estuaries from the mouths of the St. Lawrence and La Plata rivers to the Baltic Sea, Chesapeake Bay, the mouth of the Rhine, and the Bay of Fundy, are all suffering the weight of industrial and municipal effluent. These estuaries are the spawning-grounds for many ocean-going fish and are important in maintaining the productivity of the deeper ocean waters.

Besides these knowns, there are unknowns. Cylinders of nerve gas, radioactive wastes, and biological-warfare agents have been dumped

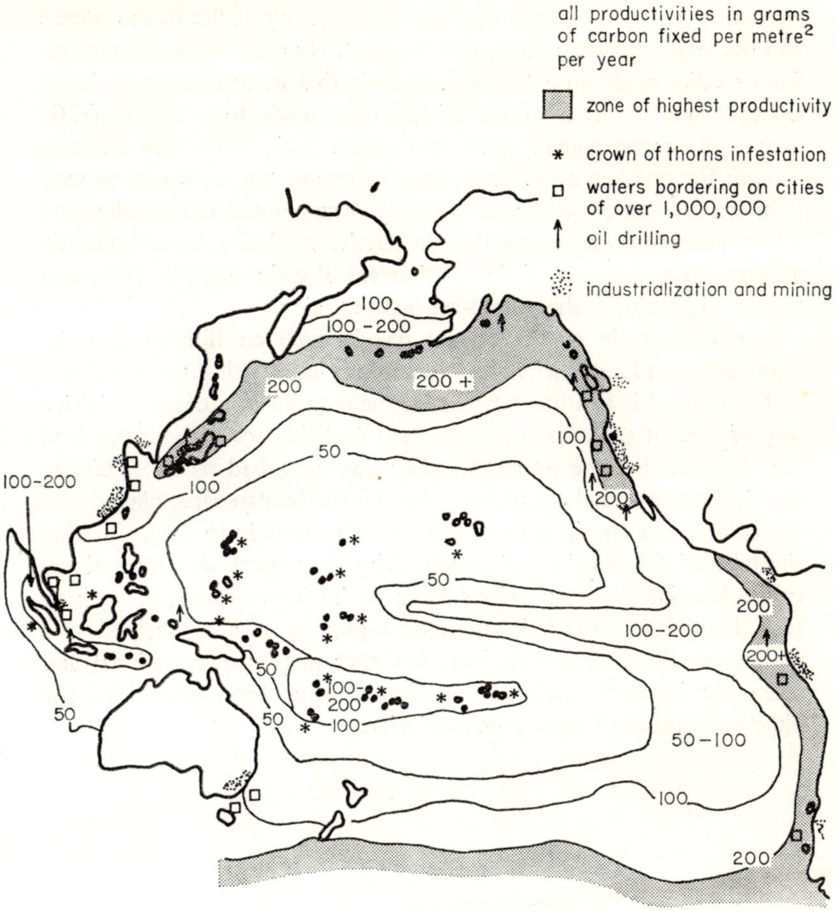

Figure 27—1 Human Impact in the Pacific

into the oceans. About these wastes there are few reports and little is known. One report has it that enough arsenic to poison the whole of the human race was dumped into the Baltic (in sealed containers) following the last war. Cylinders of anthrax pathogens have been dumped. How long these casks will withstand erosion, we can only wait and see. Last year, fishermen in the Baltic suffered mustard-gas burns from leaking cylinders that had been dropped after the last war.

On addressing the Council of Europe, meeting at Strasbourg in September of 1970, Jacques Cousteau, the eminent oceanographer, warned that in the last twenty years the intensity of life in the oceans had decreased between 30 and 50 per cent. He drew these conclusions from studies made on a 150,000-mile trip that he had just completed. Observations in these studies of life were made to a depth of 500 metres, in ocean waters about the globe. This should be warning enough for us. Apt as we have been to regard the oceans as so very rich, we are more likely to be bitterly disappointed during the next thirty years as we watch the productivity of the oceans rapidly decline. The elements of destruction are already at their work, and there will be only a short time-lag before their true meaning is felt.

The cornucopia of the oceans has become an illusion. Falsely, humans have placed their hopes in another infinity that never existed, and have used it as a justification for the continuation of an exploitive way of life. At the same time, we have foolishly squandered much of the richness that the oceans in their own way had to offer. Unless massive international efforts are taken by the human race, biologically the oceans will continue to deteriorate very rapidly to the point that shortage of fish stocks and productivity may spark an international crisis. But international efforts seem to be impossible, for we have seen how little international co-operation in whaling helped. No nation, let alone several nations, has ever put its own national and economic interests aside and made the long-term survival of the biosphere and the human race a priority.

CHAPTER 2—8

The Next Fifty Years

By the end of the sixties, in the wake of a growing environmental movement, doomsday predictions were burgeoning like an algal bloom. Modern-day prophets predicted world famine within five to fifteen years, the death of the oceans in ten years, major shortages of resources and energy within thirty years, world-wide poisoning of millions of people by pollution, and drastic changes in global temperature and climate.

The doomsday predictions, if they may be called that, are merely projections of existing trends, made to the best of present-day knowledge and under a certain set of assumptions, in much the same way that human population projections are made. They are not by any means certainties, and are fully dependent upon the assumptions that underlie them. Together, all of these predictions range in time of effect over the next five to thirty years. If they are correct, then in less than a lifetime from now, human society will be confronted with nearly all of the basic limits set by spaceship Earth. The brief time-span in which these may occur accentuates the crisis.

Of all the predictions that have been made, there are five that have withstood the intense debate of the past five years and loom ahead of us as threateningly as ever. One very real possibility has been excluded from this list, and that is thermonuclear war. We cannot, of course, eliminate it, but in comparison with the other catastrophes that may befall us, it is more dependent than anything upon the whims of human nature. Certainly as politics become more tense in the light of food, water, energy, and resource shortages, the possibility of its use is greatly increased.

Famine

Already a quarter of the human race is undernourished. Although a more equal distribution of the world's present food supply would help the situation, in order to satisfy fully requirements and to keep up with the population growth, ever-increasing food production is

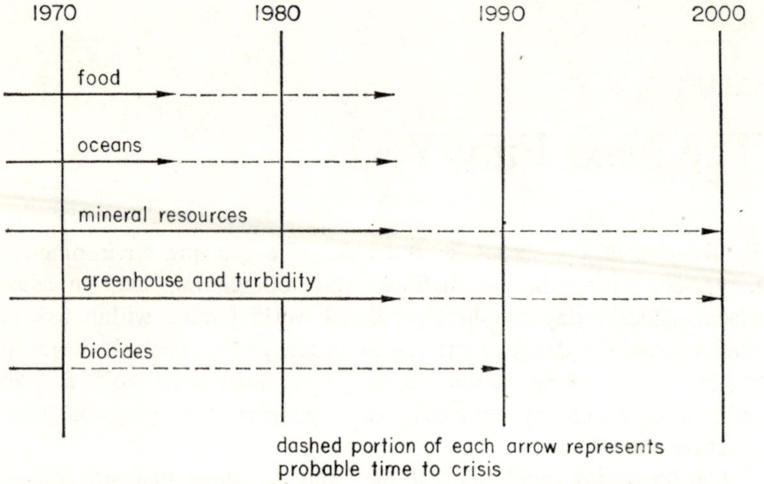

Figure 28—1 Possible World Crises, 1971-2000

necessary. During the last ten years, the developed nations—particularly the Soviet Union and eastern Europe—have managed to increase the per-capita levels of their food production. On the other hand, the underdeveloped countries have barely managed to keep these levels constant. Although in some years the per-capita production levels of the underdeveloped nations increased, in others they radically decreased. These fluctuations were not due to sudden agricultural advances, but to changes in weather that can alter total production as much as 11 per cent above or 16 per cent below the projected annual level. If India, where population growth continues at 2·4 per cent per year is taken as an example, in the twenty-year period between 1949 and 1969 food production on the average increased annually at 2·3 per cent—just below that necessary to keep up with population growth. Unless trends pick up substantially in the next five years, then the number of undernourished humans will continue to grow at the current rate of population increase. The crisis, the Paddock brothers (two U.S. agricultural experts) feel, will come about 1975. Up to now, they argue, it has been prevented by supplies of excess grain from Australia, Argentina, Canada, and particularly the United States—the only country rich enough to virtually give away food. Although the world total of undernourished has steadily climbed, rising production in the developed nations has always been able to ameliorate the situation. By 1975 however, population growth will outstrip the wheat excess of the developed world. To tighten the situation, increasing amounts of Canadian wheat are already being sold to eastern Europe, China, and the U.S.S.R., lowering the annual extra that is available to crisis countries such as India and Pakistan.

The Paddock brothers may be early in their predictions, but two things point in their favour. First of all, even if the conditions necessary for zero population growth are attained in the next few years, human population will continue to expand upwards in the next thirty. Secondly, there is certainly an upper limit on agricultural innovation—it must begin to level off. The point at which it fails to keep pace with growing population will be a crucial one.

The Oceans

The estimate of Jacques Cousteau that there has been a 40 per cent decrease in the intensity of oceanic life in the past twenty years should give us warning. Over the next ten to fifteen years, with growth of population, fishery demands, industry, and pollution, will it decrease by as much again? Over the last few decades fish yields from valued stocks have risen and then in many cases levelled off and decreased—despite the increased expenditure in energy and technology. Fishing will push towards the maximum predicted limits, but it is likely to fall short of them, for plagued with pollution and misbalance due to overexploitation of certain fish stocks, yields may again drop sharply some time between 1975 and 1985. Nations most dependent upon the oceans, such as Japan, Scandinavia, and the U.S.S.R., will, if the past is any example, respond with more intensive fishing. Subsequently, stocks will decline all the faster. For Japan, almost totally dependent upon fish protein, and also for nations such as Peru, Chile, Spain, and Portugal, the results will be devastating. Should the spread of the crown of thorns starfish not be checked, Polynesia too will begin to suffer badly.

It is impossible to predict the political results. Tensions in developed nations such as the U.S.S.R. and Japan will sharply rise as they face a mammoth protein shortage. And the protein and wheat-producing nations such as Canada and Argentina will respond by selling their excess to nations that can pay highest for them such as Japan and the Soviet Union. Conditions in the developing world, already aggravated by food shortages that could rip apart their internal unity, would break into complete turmoil and perhaps war.

Mineral Resources

If present trends continue, the shortages of mineral resources predicted by many, including the Resources and Man Committee, should be upon the developed world by about 1990. From 1980 onwards, materials such as mercury, silver, gold, and the platinoid metals will probably become short. About 1990 the shortage of any combination of copper, nickel, zinc, lead, tin, helium, and oil will place restrictions upon economic growth. Shortage of silver for photography and copper for electrical circuits may prove very important hindrances. For an unprepared society these shortages will bring about unemployment and social instability, presuming of course that between now and then the developed world does not realign its economic thinking.

About this time, some of the developing nations, for example, Iran, Chile, and Nigeria, may have reached the point of attempting to build a strong base for modern technological society, assuming that they have not been torn by famine. Their demands for resources, pitted against those of the developed world, will raise tensions. Already nations such as Libya and Chile are taking control of their resource industries, and the time will come when their exports to the developed world will be slashed. Again, either the developing nations will find resource shortages that they are unable to meet, or, they will withhold resources from the consumer giants for their own needs. In either case conditions will be aggravated in the developed world, and political dealings will toughen considerably.

Greenhouse and Turbidity

The increasing levels of both carbon dioxide and particulate matter in the atmosphere are two of the greatest unknowns that we face. We have only suggested trends to work with and an insufficient amount of evidence to make solid judgments. On one hand these factors may prove to be harmless, on the other, they may prove devastating. Hopefully, between the years 1985 and 2000, research will have devised realistic and predictable models of the atmosphere and climate so that these effects can be projected. By this time, should present trends continue, these pollutants may have effects on climate and crops, and should this prove true, then they may seriously aggravate the world food crisis.

Biocides

Research of the 1960s and early 1970s has only just begun to uncover a few effects of the many chemicals that humankind has manufactured and released. Urban smog, pesticide residues, and radio-isotopes such as krypton-85 could all have drastic effects upon human, animal, and plant life as their production accelerates with economic growth. But more frightening, and little thought of, are the effects upon the smallest and most powerful life-forms of the biosphere—fungi, bacteria, viruses, and protozoans. Already mutation rates in these organisms will have risen considerably. (Even the copiously produced molecule sulfur dioxide has been shown to increase bacterial mutation rates. Typically, this was claimed to provide evidence that perhaps mutation was induced in human beings. No one suggested any reason to worry over the bacteria.) However, with increased mutation rates among microbial organisms there are a whole range of possibilities. These pollutants may seriously affect the recycling mechanisms of soils or oceans by damaging, for example, the sulfate reducers or the nitrogen fixers. Or, through mutation they may produce potent varieties of plant and animal diseases that could sweep the biosphere and the human race. Or, they could produce beneficial varieties. The last alternative, however, is the least probable, since genetic evidence has shown us that the majority of mutations are either harmless or harmful, but seldom beneficial. This fifth danger signal could be the most potent of all, yet it is the least predictable.

Conclusions

In 1970, Jay W. Forrester, a systems-analyst at the Massachusetts Institute of Technology, uncovered work he had carried out on a computer model of the global environment and the major human forces within it. These he deemed to be: population growth, pollution, availability of natural resources (including food), economic growth, and the quality of human life. These factors were interrelated closely with his model system. His conclusions are striking and tend to reaffirm those reached in this book. Only under a very radical set of circumstances did he see the possibility of easing the world situation and reaching a new balance. These circumstances included cutting population growth, resource consumption, economic growth, and pollution.

If population alone was to be reduced, increasing economic growth would bring the human race up against resource shortage and a collapse. Should pollution alone be curbed, human population would rise drastically through better living conditions only to crash downwards from resource shortage. Should capital investment in environmental quality be increased, it would extend human growth a short distance beyond the limits of pollution to meet head-on again the limit of natural resources. And should capital investment and population be reduced, it would only forestall the resource crisis for a little longer. Finally, even if resource consumption were cut, increasing population growth would eventually overcome this and increased pollution and resource shortage would bring about the same end. Forrester warns urgently against partial solutions to the environmental crisis and suggests that all these factors must be accounted for.

As a result of his studies, Forrester comes to a number of highly relevant conclusions. Industrialization, he predicts, may have a more potent effect upon the biosphere than population growth, since the weight that a single person in the developed world places upon the environment is twenty to fifty times greater than that of a person in underdeveloped nations. Furthermore, societies at a high level of industrialization appear to be incompatible with the limited nature of resources and may be inherently unstable phenomena. Perhaps, he says, the efforts of the underdeveloped nations to emulate Western patterns of economic growth may be unwise. Not only might they be nearer to a state of equilibrium now, but being less dependent upon technology they will have less of a fall during a world collapse. Finally, Forrester thinks that, on the average, humanity today may be living at a higher quality of life than ever in the past or in the future.

Forrester's results, and he would certainly agree, are not all conclusive. They merely study the various interactions of human forces. More than anything they point out how closely the fundamental roots of the environmental problem—energy, resources, population, and the biosphere are tied together. His estimates to time of disaster are, on the average, fifty years later than the ones implied in this chapter. Perhaps this is because a computer model, even as complex as the one quoted here, cannot account for all of the biological details involved in the problem. (The model, for example, appears to have considered the effects of pollutants only upon the human race.) Hopefully, Forrester's predictions are more accurate, for, if so, it will give us more time for the massive and complex changes in human systems that are necessary for survival.

PART THREE

Earth Is All We Have

'The force that through the green fuse drives the flower
Drives my green age; that blasts the roots of trees
Is my destroyer.'

Dylan Thomas

CHAPTER 3—1

Dynamic Equilibrium : A Theme for Human Survival

Evolution and Ecology

Animals, plants, and bacteria grow, reproduce, die, and are decomposed. Together their individual life-cycles form the workings of the many ecosystems about the globe, from grasslands to coral reefs. Together these ecosystems and the interactions between them that extend from the deepest trenches of the oceans to the highest mountains constitute the biosphere. At any point in the flow of time, the biosphere is in ecological equilibrium. Through the food chains from the producers to the decomposers, living things are interacting in a balanced way. But through time the equilibrium is changing. In response to long-term shifts of climate and geology, evolution is sculpting the features of life as seen on a species level. In the short term, in response to sudden fluctuations of climate and geology, equilibrium is re-established through the process of succession.

Think of a water barrel that has an opening at the top and an outlet at the bottom. Water poured in the top will flow out the bottom. However, when the rate of water entering the barrel equals the rate of water leaving it, then dynamic equilibrium has been attained—the water level remains the same even though the water itself is moving. The biosphere thus functions as a dynamic equilibrium—one that is always changing and moving yet is always in a state of balance.

This equilibrium is achieved through four fundamental principles. First of all, energy from the sun is the driving force of life, and as it is used it is irretrievably lost as waste heat; thus it cascades through the biosphere. Along the way it is used to manipulate the matter needed by all living things.

Secondly, matter is cycled about the biosphere, used again and again so that there is no such thing as waste. Thus, the nutrients necessary to life are always available, and the toxic by-products of many living organisms are metabolized back to nutrient material.

Thirdly, within an ecosystem matter is flowing through the living organisms of each trophic level, and population-control mechanisms are at work to regulate this flow. Thus the inherent stability or persistence of each ecosystem—marsh, prairie, taiga, or reef—is linked directly to the stability of the biosphere. And on an ecosystem level, this persistence level is related to the diversity of the system.

Further, through time, all ecosystems are striving towards a climax stage which is characterized by a production/respiration ratio equalling one, and a high diversity, stability, and information content relative to all of the stages before it. At this climax stage biological control over the environment is maximized. Some ecosystems may not reach this point because factors such as rainfall or nutrients may be limiting. They do, however, reach a stage of maximum stability under the prevalent conditions.

The stability of the biosphere appears related to the complex feedback mechanisms that work within and between the organisms of an ecosystem. In other words, the response of one organism to the stimulus of another may alter the initial stimulus. Furthermore, dependent upon the intensity of the stimulus and a large number of other variables at any given moment, the response can take many different forms. Thus, the lynx population responds to the population of its prey—the hare—and vice versa. Should the conditions at the bottom of a lake become anaerobic, then the sulfate-reducing bacteria may set to work; should they remain aerobic then a completely different set of processes will take place. Although plant communities respond initially to the indigenous climatic and geological conditions, over time they may modify these, so that the plant communities will further shift their responses.

It is because of feedback responses within ecosystems and within the biosphere as a whole that stability can be maintained. And yet it is because of these responses that the system is so difficult to understand. We must not become overconfident with our present knowledge of ecology, for although science has managed to map out some of the over-all trends, there is very much that we do not know. At every level of the biosphere minor peculiarities exist that are of utmost importance to an understanding of the system. Because even a small organism is capable of responding in complex ways, the construction of a working model of a simple two or three organism system is a very difficult task. Populations of thousands of organisms participate in the pathways of a natural ecosystem.

There is little doubt in the minds of scientists today that a great part of the stability within the present biosphere is derived from millions of years of adaption, adjustment, and co-evolution of living organisms. But again, we know too few of the details of evolution and have only been able to map out some of the broadest trends. We know hardly anything of the shifting roles of various organisms both within individual ecosystems or within the biosphere as a whole, and we know even less of their evolutionary impact. We are only just beginning to speculate upon some values for organisms such as the nitrate and sulfate-reducing bacteria, or the photosynthetic algae, but these numbers are crude and reflect our basic ignorance of quantitative ecology.

The Forces of Change in Human and Biological Activity

For the most part, changes within the biological world have been long-term ones. More than three billion years of biochemical evolution gave rise to the first self-reproducing organisms and to complex cellular organs such as the mitochondria and the chloroplast. The oxygen revolution took millions of years to bring about, as did many of the effects of biogeochemical cycling. The processes of selection and diversification took many millions of years, and even the lifetime of a single species between speciation and extinction averages 600,000 years for a mammal and 2 million years for a bird. Climatic changes, too, are a long-term phenomenon—geological changes such as glaciation and warming and cooling of the planet take hundreds of thousands of years. On a community scale, too, changes take long periods of time (in human terms), and processes of plant-community succession and soil building span many centuries.

Against this backdrop the changes wrought by human cultural evolution are sudden ones. Starting with fire and advanced tools, primitive humans brought about through the Pleistocene over-kill, in a period of 100,000 years, irrevocable changes in the structures of ecosystems about the globe. Agriculture was a more sudden process, its effects being felt in a time period only about one-tenth of the Pleistocene over-kill. Following its introduction 10,000 years ago, after about 8,000 years vast portions of the planet had lost their soils and climax communities, and in some cases had lost the ability to regenerate them.

Human-induced changes appear to be self-accelerating. One technological innovation leads to another. Primitive metallurgy led to better tools and weapons, the wheel, and simple machines, all in logical progression. The discovery that one metal could be extracted from rock most certainly led to a search for better metals from other types of rocks; in other words, the discovery of the first metal was an essential step to the whole concept of metal extraction and use, a step that accelerated further discoveries. Compare, for example, the extent of human knowledge and of technological innovation between the agricultural revolution and the Roman Empire, between the Roman Empire and Reformation Europe, and between Reformation Europe and today. The growth of technology has been an exponential one, and the effects of technology upon the biological and the human environment have grown in that manner. Greater changes have been wrought in less and less time as the curve of technological growth became steeper.

Technology continues to grow. This is reflected in the Western world in the number of new drugs and appliances reaching the market every year. In the period of the next thirty years the power industries expect to move from fission to fusion reactors, yet it took a period of 500 years to move from coal to fission. One cannot help but ask, do we have control over our technological growth? Or does it have control over us? Are we so infatuated with the 'ideal of progress' that we cannot slow down technological innovation in order to improve the quality of the technology that we already have? But so obsessed have we become with technology that we hold it as the answer to all of our problems, past, present, or future. As Ontario Hydro puts it so succinctly: We cannot expect man to curtail his activities, therefore we must turn to technology for the answer to environmental pollution problems. To the proponents of technological utopias, like Ontario Hydro and the Atomic Energy Commission of the United States, all of the basic limitations of spaceship Earth will be overcome by newer, better, and of course, pollution-free technologies.

Ignoring the recent explosion of technological growth, technology has still been a sudden force upon the biosphere. For the most part it has occurred within the last 100,000 years and to a much greater extent within the last 10,000 (Figure 31—1)—in direct contrast to the hundreds-of-thousand-year periods in which evolutionary and geological processes tend to operate. For these reasons human cultural evolution is probably not a normal biological process and is a very different phenomenon from the biological evolution that preceded it, even though the former may have had its origins in biological

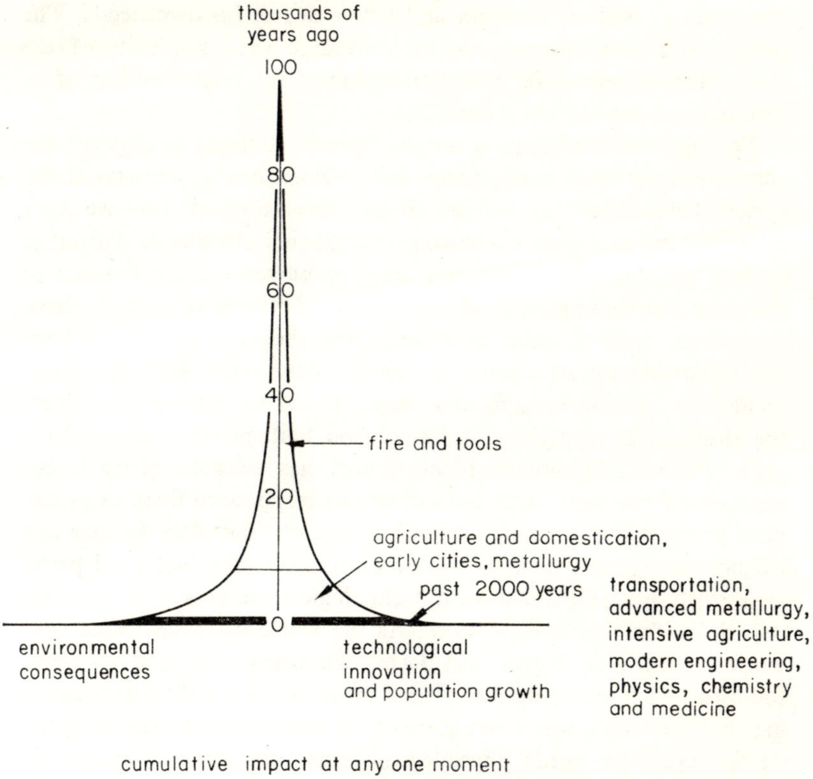

Figure 31—1 Technological Growth

mechanisms. Apart from the impact in time, there is another basic difference between the processes of the biosphere and the human use of technology.

On the one hand, ecological systems tend to be exceedingly complex ones in which it is difficult to understand and to predict the details of events. On the other, humans have what could be termed a 'stock-inventory view' of the world, in which every object is classed as either useful or useless for human purposes or technology. It is best exemplified in our conception of the one-way waste-production arrow, where we conveniently define materials useful to us as products and the others as wastes. The technological process is a simplified approach, analagous perhaps to an individual who, ignorant of all but a glimmering of electronics, approaches a modern computer with the intent of making it bigger and better to suit his own needs. The project is almost certainly doomed, because the complexity of the system forbids successful alteration without a full understanding of its features and the way they work.

Through the brief span of human history, attempts to improve the computer have been made many times. Consequently, as parts of the system have failed—as various circuits have blacked out—we have attempted to save face by making adjustments elsewhere. To put it another way, we have time and again employed our intelligence to postpone the consequences of our actions. Agricultural systems have washed the soils of stable ecosystems into the seas, but people have not responded by attempting to rebuild those soils—they have responded by moving on into new areas. Today we attempt to correct the situation by applying fertilizers. And both of these moves have other effects. The use of monocultures has released many insect organisms from their natural controls; our response to these pests has been to apply indiscriminate chemical controls that often further unbalance the system. Even to what are fundamentally biological problems, man has responded with technological answers that have no regard for the processes at work. As the whale catch decreased, the whalers turned to bigger and faster whaling-vessels and to more sophisticated tracking devices. Were these men incapable of conceiving that the whale was a living creature, that the whale had a reproductive cycle set within physiological limits, and that no amount of technological gee-gaws would ever make the whale grow or breed any faster? Even today, in the light of depleted fish stocks, the Food and Agricultural Organization of the United Nations has expounded the need for bigger and faster fishing-vessels and a greater technological advantage over the fish (despite the urgent warnings of many of their own biologists that better-regulated and more evenly distributed fishing is required.) The human race lives in a biological world that is extremely complex; this has yet to be understood. Technology has provided us with a very successful way of avoiding many of our mistakes—so far. But the mistakes have accumulated behind us, and perhaps the intensity of this accumulation has something to do with the recent rapidity of technological growth, as we try harder and harder to improve a situation that is deteriorating more rapidly at every moment, and attempt to do so with methods that only make the situation worse.

Technology has done much for the human race, and it must serve an important role in helping us out of our present problems. Technology is essential to the state of modern society today, and it is only through technological modifications made on existing technology that we will be able to alter the conflicts between human society and the biosphere. These corrective measures must not, however, be applied with the same simple-minded approach that has characterized nearly all technologies of the past. The required approach must comprehend the biological nature of the problems it is dealing with, and must search for new frameworks within the mechanisms of the biosphere to employ technology. We must therefore determine the limits of the planet to support human activity. Within these limits we must operate using the principles by which the biosphere operates. Matter must be recycled and energy used in ways that fit the cascade of energy through the biosphere. Biological diversity must be maximized and biological patterns of climax and succession encouraged. Technology must therefore be geared to ecological systems, and decision-making processes must reflect, as ecological processes do, the totality of life, earth, and the biosphere. Long-term stability must be balanced against short-term effectiveness.

The state of human consciousness and memory are perhaps ill-equipped to grasp the long-term nature of biological change. Human beings have operated best on a day-to-day or year-to-year basis, and this life-style would have been quite adequate if it were not for the sudden and far-reaching consequences of technology. In a period of years changes are wrought that previously might have taken thousands, or might never have occurred at all. The extinction of species, the simplification and modification of gene pools, the introduction of new species to ecosystems that are unadapted to them, the consumption of non-renewable resources and energy and of soils and forests, the pollution of watersheds, and the contamination of the biosphere with carbon dioxide, long-life pesticides, and radio-isotopes, all of these phenomena occur rapidly within periods of years or decades. They will affect the biosphere for centuries and perhaps even for millenia, and the course of history as long as people inhabit this planet. Yet they are measured in terms of immediate losses or gains, mostly economic ones that are accrued by a few individuals. Irreplaceable heritages of all people for all time, non-renewable resources are consumed by industry and government as if there were no limit on their availability and no generation of humans after this one. Consumption is accelerated by overselling, wastage, and inefficiency—in the Western world this is essential to keep the G.N.P. rising. And the potential needs of future peoples are rationalized by saying again: technology will find an answer. Human beings must, therefore, learn to judge themselves as part of what is a long-term biological continuum, and all actions must be measured against this continuum, as well as against the immediate benefits to human health and well-being.

Yet we have still to learn how to make decisions in terms of human well-being, for today the majority of the human race lives in abhorrent conditions, while a small minority continue to justify their own existence in terms of maintaining human well-being. Environmental problems are simply one aspect of the many sociological and cultural difficulties that have accumulated in the human world over many thousands of years. Like many of the environmental problems, they have their roots in the simplified and egocentric way in which humans have viewed themselves and the biosphere. Today, like the environmental crisis, they face us with great urgency. In many cases, they too can no longer be postponed. And as we have seen, nor are these problems of life-style and environment unrelated.

In total, we must attempt to come to a state of dynamic equilibrium with the biosphere by finding the means to integrate a human way of life with the biological mechanisms that preserve our environment. To do this will require long-term and extensive planning of the sort to which individuals and nations have not been accustomed. It will require new directions for all of the major technological and social processes, and it will require a new look at both our political and our economic systems and their priorities. But against this background of long-term planning on which the human race must embark immediately, sporadic and short-term attempts to rectify the situation must continue. We must continue to enact and to enforce legislation against pollution, we must plan against the cancerous sprawl of urban growth, and we must accommodate for the growth of population which is inevitable between now and the year 2000, even if we should immediately institute the means necessary to attain zero population growth.

The short- and the long-term approaches must work hand in hand, for without both the human cause is hopeless. If we attempt to improve the human condition by short-term solutions alone, we will only run into insurmountable problems as we outgrow the carrying capacity of the biosphere and come hard up against the shortcomings of solutions that do not consider in the fullest light the integrity of biological systems. On the other hand, some of the most potent poisons are entering the biosphere from today's industry, and some of the last wilderness areas are being invaded and turned under. If we plan only for the long term and neglect the present state of affairs, then we may never reach the long term, or we may get there having nothing left to live with.

Zero Population Growth : New Controls for Old Controls

Through cultural evolution we have released many of the population controls that existed once during human life in a Pleistocene ecosystem. More recently, medical technology has dramatically released the controls of disease, and the human population has exploded upwards in the last hundred years. It is now booming upwards with a global doubling-time of under thirty-five years, less than the average human lifetime today. Almost a billion of our people are undernourished, and there is a great disparity in conventional standards of health and welfare and in the consumption of resources and energy among the world's people. The prospects of satisfying these needs do not appear promising; rather, they look dismal. We are thankful in today's world that modern medicine has averted massive suffering, but at the same time we must be aware of the great potential suffering that medicine may bring through human famine, plague, and underdevelopment.

If we have any desire to control this situation, then we have two fundamental choices. Either we can drop the advances of modern medicine, or we can apply new controls to replace the ecological ones that we have dropped. If we chose to drop modern medicine, then we chose to lay aside much that human cultural evolution has achieved in enhancing the human condition. Therefore, we must control human population size, and we must control it through the application of medical technology in another way—through programmes of birth control.

The programme must be one of population control rather than of family planning. Surveys show that on a global basis people have about as many children as they desire: in the developed nations, on the average, a few less, and in underdeveloped nations a few more. These desired family sizes are related to indigenous social and economic conditions. In the developed nations the cost of clothes, social services, and education seem to be influences. In the underdeveloped nations the need for children to work in the fields and support the parents during their old age both appear to be important factors. Only in a few countries does the average number of desired children lead to a zero growth rate, excluding the influence of migration. Broad public education on the global and environmental necessity of population control is, therefore, an essential part of the problem.

As an essential part of this programme, mechanisms for the prevention of unwanted births must be available to all members of society. Free and unimpeded access to birth control, vasectomy, and abortion are essential rights of all human beings. Today, but for a few nations such as Great Britain and Sweden, this is hardly the case. In Communist countries, availability is linked to government policy—whether an increase or decrease of population is desired. In Catholic countries such as France, Ireland, and Italy, all of these services are difficult to get, or are unavailable, except to the rich. In Canada and the United States legislation is often prohibitive, especially with respect to abortion. In the underdeveloped nations, shortage of communications systems and trained personnel are problems; especially in Catholic countries, legislation may be prohibitive.

Population control can take many forms, dependent upon necessity. It does not necessarily imply an immediate and a rigid curtailment of human freedoms. In many cases the mere formulating of a national policy based upon zero population growth and coupled with a broad educative programme on the social and environmental necessity of this, could lead to the desired growth-rate. If such a programme failed to bring about the desired result, economic incentives could be used. In Canada, as in many parts of the world today, large families are encouraged through a series of these incentives. Income-tax deductions increase with additional children, and baby-bonus allowances are distributed to parents of children. These incentives could be formulated to work in a reverse manner, to encourage, on the average, a family size of two children per couple. Finally, should these means fail, governments may have to resort to methods of greater severity, including the legislation of family size, compulsory birth control, or compulsory sterilization of any male having produced two children. These measures, however, are only the extreme case should all others fail. The largest part of the problem is educative and lies in convincing the public of the environmental crisis that we now face.

A population-control programme is not a restriction of human freedom. Like any human action, the raising of children must be weighed not only in immediate terms, but also in terms of the future. The absolute freedom to produce as many children as a couple desires today will inevitably lead to a loss of freedom among their children and their children's children. For as human population density rises, so do the social restrictions that are necessary to preserve order and environmental quality. And the smaller availability of resources, energy, and free space will severely limit the activity of individuals in the future.

What is needed, perhaps, in today's world of spaceship Earth, is a new attitude towards children. Every individual partakes of the food, resources, and energy that are available on this planet. Every human individual is part of a vast and beautiful living system. He or she has a right to be wanted by that system and loved by other humans in that system. The preciousness of life requires that we only produce children that can be guaranteed a fair chance of love and interaction with both the human and the biological world. In the search for a cultural value to life, the human race has removed the indiscriminate controls of an ecosystem. But to retain the cultural values that we have achieved, controls are now necessary. These we must impose ourselves.

On a global basis the impetus for zero population growth must come from the developed nations, and it must come first from within their borders. In the light of the pressure that North Americans exert upon the world's resources and the biosphere, there is little doubt that they have reached the level of saturation. In the past five years, both Presidents Nixon and Johnson have commented on the crisis of exploding population, and their government have gone to the extent of including family planning in their foreign-aid programmes. They have yet to tackle the problem at home, however, and to devise a similar programme of population education and control for the participation of *all* citizens. So far, any governmental programmes that have been introduced in America appear to favour population control within certain social groups—such as poor blacks—while not in others. Programmes like these stink of genocide and have no place in population control. Nor is there any place for programmes of minorities to radically increase their numbers by breeding in order to gain powers. In the Western world the population explosion stems from all rural, urban, and suburban growth alike, and an adequate programme must reach all social groups. Furthermore, until Western nations institute their own comprehensive policies of population control, it is unfair of them to demand that other peoples do so.

Optimum Population

We must determine the optimum population for Earth. To do so will require an inventory of the renewable and non-renewable resources, an extensive summary of the strains that human activity places upon the biosphere, and a view of the long-term survival and health of the human race and of the biosphere. The Resources and Man Committee (of the United States National Academy of Sciences, 1969) argues that although the world could probably maintain fifteen billion people at a level of *impoverished existence and near chronic starvation*, a population of seven billion is the maximum that could exist at a fairly adequate level. For the long-term stability of the biosphere, the availability of natural resources, and the health of the human race, the committee feels that a population *less than the present one* would be optimum. Although these estimates are likely to be crude approximations, they do give us an idea of where we stand today—probably over the optimum and only thirty years from the seven billion mark. With a sense of great urgency then, we must take the crucial steps towards determining optimum population size, and then bringing it about. Some of the factors to be considered are the following:

1. *AGRICULTURAL LAND* What is the capacity of Earth to produce food on a sustained-yield basis? In making such a calculation the long-term effects of agriculture must be truly considered.
2. *MINERAL RESOURCES* What are the probable and recoverable reserves of industrially important minerals? What are the environmental costs of extracting these?
3. *ENERGY RESOURCES* What is the total potential of energy resources available to us on this planet? What is the maximum population that we can support on the cleanest methods—hydro, geothermal, and nuclear, and how much environmental disruption from these processes are we willing to permit?
4. *RENEWABLE RESOURCES* What is the capacity of the biosphere to provide us, on a sustained basis, with materials such as wood, hides, and plant oils?
5. *TECHNOLOGY* How much, within reason, can technology be expected to alter the above predictions? What are the inherent dan-

gers of overreliance upon technology?

6.*PER-CAPITA CONSUMPTION* What are reasonable per-capita consumption levels for food, water, energy, renewable, and non-renewable resources? What sort of buffer zone must be allowed about these levels in order to allow for fluctuations in use at the consumer level, technological innovation, and new products and processes? (Incidentally, per-capita levels include not just personal consumption but the material spent in a society on research, education, and maintenance of public facilities.)

7. *RECYCLING* How much will total recycling affect the previous consideration and thus the total demands of human society on the resource base?

8. *WILDERNESS* What is the per-capita demand for wilderness, on the basis of that needed for ecological stability and on the basis of recreational needs? Can these wilderness areas, at least in part, be rotated over fifteen- to fifty-year periods with land used for agriculture and forestry?

Using criteria such as these it should be possible to set an upper limit on human population size. There is also a necessary lower limit and this is determinable; it is set by the size of population necessary to maximize genetic diversity and to operate the institutions of modern society. Within this envelope of maximum and minimum size, human population must be optimized. And should this be achieved, the human future will hold great promise, for we shall have attained a world in which human gratification and dignity are possible at the same time and on a global basis.

Agriculture and Sewage Treatment

Farming is an old technology and in many parts of the world its methods are rooted in tradition and will prove difficult to change. Because agriculture provides us with our daily bread, we can only continue in our present ways, hoping to slowly implement sounder methods.

These new methods span many fields. They must be aimed primarily at reproducing within a crop community the mechanisms that preserve long-term balance within a functioning ecosystem. We should, for example, be building up and maintaining topsoil and decomposer systems with composted organic matter and human and other animal excrement. We should be experimenting with crop systems that are polycultures rather than monocultures, and that provide adequate vegetative and root cover to the soils. There are many possibilities of devising sounder irrigation schemes, where networks of plant roots—including the crop—retain water and soil particles and also slow down the erosion that so often accompanies irrigation. In total, we must find methods of agriculture that, although maximizing the human-oriented production, also maintain the integrity of a stable and functioning ecological community.

One highly successful approach of this type is that of integrated pest control. Realizing that a pest is simply a phase of a complex system that is out of balance and that can react in many ways to indiscriminate attempts to stabilize it, one must search for a factor that will predominantly affect the pest organism alone. Primarily, a good understanding of the pest, its habits, and its relationship with the crop ecosystem are needed, and these must be researched. With this information there are a number of possible approaches. One may release a predator of the pest, perhaps a wasp or an insect virus, onto the crop. The value of these controls is significant. Not only do they remain from year to year, but the insect will not gain resistance to them in the same way that it can develop resistance to a pesticide. Seventy years ago, scale insects threatened to destroy the California citrus industry. The vedalia beetle, a ladybird predator of these pests was introduced; today the scale insects are still under control, with no additional treatment necessary. Other methods may also be used. Large numbers of the pest may be bred in captivity and sterilized with ionizing radiation. Then, during mating season, they may be released, in order to swamp the native population with sterile partners. This programme has worked very successfully in the southern United States against the screw-worm fly, a cattle parasite. Or, chemosterilants may be mixed with synthetically produced insect attractants in trays at field edges. The chemical characterization of these potent sexual attractants has made it possible to attract specific insects, particularly during mating season. Sound and light traps are also possible. In total, the integrated approach requires extensive research before it is fully practical, but in the end the value of the investment is far above that put into pesticides. In California, four to five dollars were made on every dollar invested in chemical controls, but on biological controls the returns were thirty dollars for every one dollar invested. If ecological value were computed into these figures, integrated control systems would be way ahead. It is a sorrowful situation when the governments of the developed world, because of pressures from the chemical manufacturers, are emphasizing chemical solutions to biological problems.

On a world scale, more research into the life-cycles, habits, and ecological relationships of tropical pests and their predators is needed. Pesticide usage is very high in the developing world and is virtually the only control applied against the malarial mosquito. Crucial research work must be put into integrated control systems for the mosquito and for the tse-tse fly.

There is one new direction in which agriculture must turn, and this is to reconcile the unbalanced phosphate cycle. We must find ways of slowing down the loss of phosphate from the land masses, instead of accelerating it. Methods must therefore be found to return sewage and biological wastes to the land in a manner that stimulates soil building and plant growth.

Conventionally, these wastes have been treated in a disjointed manner, as the problems associated with the release of raw sewage into the environment became evident. Steps such as settling, to remove suspended solids; chlorination, to kill pathogenic organisms; activated sludge treatment, to decompose materials with a high oxygen demand; aeration, to restore dissolved oxygen; and even lime treatment in order to remove phosphates were all added one upon the

Figure 31—2 A Biological Scheme for Sewage Treatment

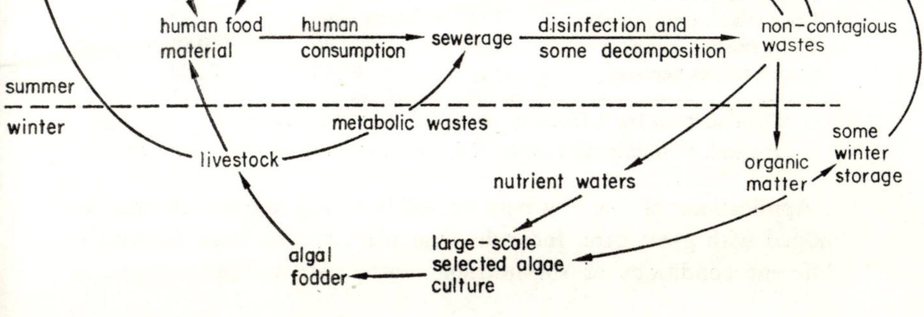

other, with no overview of the very biological importance of sewage and waste organic matter. Notably, the costs of such a process become more and more expensive as primary and secondary treatment plants are augmented with tertiary facilities to remove phosphates. And this method produces large quantities of insoluble phosphate and sewage sludge. At the other end of the balance, the nutrient content of the soils is decreasing with the simplicity of monocultures and the diminishing returns of organic wastes.

Figure 31—2 outlines an ecological approach to this problem. Sewage is treated to remove infectious organisms; this step is essential. In China and other Asiatic nations, where sewage has been returned to the fields for many centuries, nearly all of the rural population is infected with parasites such as the tapeworm and the liver-fluke. (This has certainly been intensified by the density of the Chinese population.)

The sewage is then partially decomposed to attain some degree of homogenity and to release some of the soluble nutrients. Nutrient-rich waters are then separated from the solids. Both are returned to agricultural and forested areas. The waters may be used for irrigation and the solids may be further treated to produce a soil additive.

In total, this procedure would involve an expenditure of energy that, in the case of a largely urbanized population, might be high. Costs could be minimized by stimulating the growth of small cities separated by agricultural districts. The results on crop yields would be significant. Experiments in the U.S.A. have obtained crop yields far above those obtained with the use of conventional fertilizers, in some cases doubling them, as the following table indicates.

Table 31—1 Crop Yields at Various Levels of Waste-Water Application*

(inches per week)	0†	1	2
Alfalfa hay, tons per acre	2·27	4·67	5·42
Corn silage, tons per acre	3·11	3·93	4·32
Corn, bushels per acre	63·30	114·40	110·80
Oats, bushels per acre	45·20	80·10	72·60

*Data adapted from *Environment*, Vol. 12 (2), 1970.

†Instead, recommended doses of fertilizer for that area were applied.

Applications of waste waters or solids to any ecosystem must be judged with great care, for individual plant species have adapted to different conditions of soil-nutrient concentration. For example, in the project carried out above, red-pine growth was found hindered slightly by the application of waste waters, while annual increases in young oak trees were 27 to 83 per cent greater than those in a control area. In systems of conventional agriculture, nutrient loss appears to be accelerated and even applied fertilizers are quickly washed away. In contrast to these experiences, a large percentage of the phosphate, potassium, and nitrate in the waste waters was absorbed by the plants during the experiments conducted.

In regions where there is seasonal freezing and dormancy of plant growth and bacterial action, an alternate route for nutrient flow is necessary. One is suggested in the lower half of Figure 31—2. Sewage treatment would be carried to a stage of partial decomposition as before; the semi-decomposed solids would then be separated from the nutrient-rich liquids and stored, perhaps by freezing, for further use during the growing-season. The nutrient-rich waters would then be used as a base for mass algal cultures, as has been done by the Japanese. Algae chosen for their productivity and nutritional value could be harvested and used as livestock fodder. Great care would have to be exercised during the culturing of the algae, to prevent contamination by poisonous blue-green varieties.

A New Economics

The economic systems of today's world are based upon two major misconceptions. One is the super-abundance of natural resources; the other is the necessity for a continual growth in resource (and energy) consumption that is subsequently reflected in the G.N.P. Both of these concepts are at direct odds with the finite nature of Earth and the biosphere and the sustained patterns of resource consumption that characterize ecosystems.

It is imperative that we develop a new system of economics in which to manage human well-being and the use of resources and energy. Such a system could follow the steady-state patterns of the biosphere. Thus, resource uptake will remain at a constant rate, sustainable by the availability of natural resources. Within the human sphere, the use of any resource will be maximized by recycling. As a result, the annual demands upon the resource base will be minimized and what is consumed annually is only that needed to replace losses due to corrosion and wear within the system. In a steady-state economic system, resources such as copper would be a highly crucial factor, for unlike paper and other biologically derived materials, they are only renewed by very slow geological processes.

A steady-state economic system does not preclude competition or diversification, both of which are possible beneath a maximum limit of resource consumption. Such a system does imply a much tighter economy than we presently operate, and leaves no room for extravagant resource usage, a large part of which has gone solely to sustain the concept of growth. Present trends of planned obsolescence and gadgetry are irresponsible directions of resource consumption that will not be tolerated. At the same time, the individual benefits to be gained through wise and purposeful use of resources will be immense.

A steady-state system of economics cannot and should not be applied immediately and universally to all national economics today. Rather, it must be an ideal towards which national policies will evolve over the next ten to one hundred years. As Forrester's work implies (see Chapter 2—8), the adoption of a steady-state economy is more nearly imminent in the developed world where resource consumption is so disproportionate. The developing nations must integrate a steady-state future into their programmes of economic evolution. And needless to say, on a global basis steady-state economics must be accompanied by zero population growth.

Resource Management and Industry

In any use of resources we must come to grips with two very fundamental principles.

First of all, resources are limited in obtainable quantity on this planet. The ultimate yield of a particular resource may be altered to a certain degree by technology, yet an upper limit on its availability does exist. Furthermore, the more concentrated a resource, the more valuable it is to us. Modern technology, through the waste production arrow is dispersing valuable resources over the surface of the earth. Surely we should be doing the opposite, maximizing the long-term value of these by maximizing their concentration through recycling. Then there are limits on the availability of the various forms of a resource. Many types of wood, for example, may be used for making paper. However, the limited yields of fine woods such as mahogany and yellow birch preclude their use in this manner.

Second, matter flows through the great biochemical cycles by the action of life, and these cycles are essential to the short- and the long-term stability of the environment. Industry must learn to work within these cycles, or else to imitate them. In the case of materials that are derived from the biosphere (such as paper or oxygen), technological systems must be constructed into the natural cycles in order to maintain the balance. In the case of materials that are derived from without the biosphere (such as nickel, cadmium, or uranium), cyclic processes that are closed from the biosphere must be devised.

Let us consider a few applications of these principles.

(a) Global Resource Management

This over-all approach is based upon the availability of resources in their various forms, and the demands of human industry for them. We will briefly examine this sort of approach using carbon resources as a model.

Carbon resources consist mainly of four forms of carbon:

(i) Plant and animal carbon—for example, wood, hides, and fibres, plant and animal oils, and rubber. These may be processed into materials such as paper, clothstuffs, and soaps. All of this carbon is essentially part of the biosphere.

(ii) Fossil-fuel carbon—for example coal, oil, gas, and shale oils. This carbon is in a form uniquely different from the carbon in life; it has been transformed into a mixture of alkane and aromatic hydrocarbons. It may be chemically transformed to plastics, synthetic fibres, petro-chemicals, pharmaceuticals, or used directly as fuel for heating, electric power, or transportation.

(iii) Limestone carbon. This is the carbonate that makes up limestone rocks. It can be used for construction materials, in which case it generally remains as limestone, or it may be heated to produce lime, which is employed in building or smelting. In this case carbon dioxide is released to the atmosphere.

(iv) Atmospheric carbon dioxide. Used for few industrial processes, this form of carbon is occasionally used in a changed state—either dissolved in water or as dry ice.

The use of carbon products must therefore draw together the availability of carbon in these forms and the flow of carbon in the biosphere. Figure 31—3 illustrates this. Two important consequences appear. First of all the *ultimate end-product of human carbon-use is almost always carbon dioxide.* All of the biological carbon products —wood, paper, cotton, and so on—usually end their pathway by incineration or by decomposition. The qualitative entry of this carbon dioxide is not an unnatural one, for should this carbon have remained unused by industry, it would have faced a similar end. It may not,

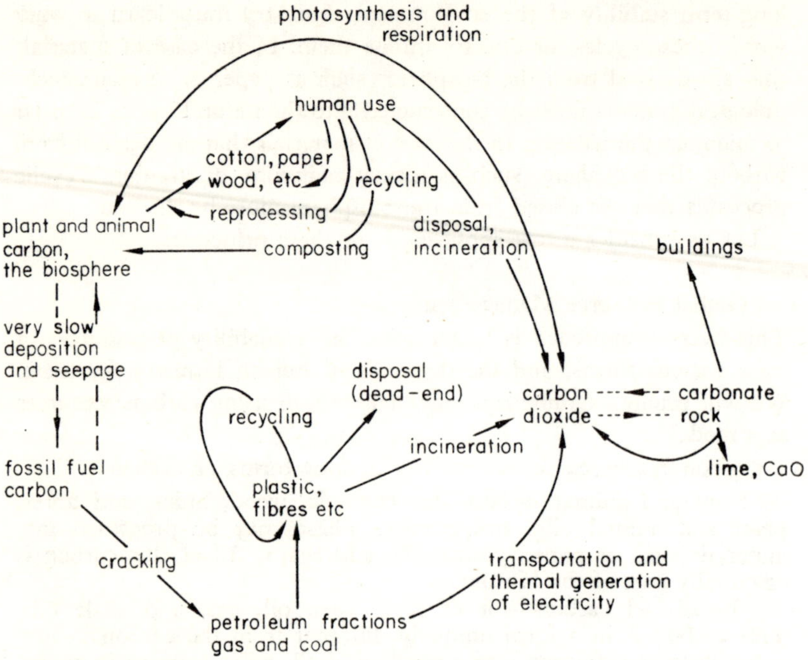

Figure 31—3 Carbon Resource Management

however, have decomposed at the same rate. Fossil fuels usually end up as carbon dioxide as well, either through burning as fuel or by the incineration of plastics and artificial fibres. Both of these sources are an 'unnatural' entry of carbon dioxide into the biosphere.

The crux of this management problem lies in rates. If the rate of entry of carbon dioxide into the atmosphere were so small in comparison with any of the natural inputs (for example, respiration or erosion of carbonate rocks) that it was negligible, then as a factor of importance in resource management it might be ignored (as long as it caused no other harm). However, the rate of entry of carbon dioxide into the atmosphere is not negligible, for it has noticeably increased in concentration and is continuing to do so. It must, therefore, be considered as a problem.

The second fundamental lies in the type of carbon resources that we employ. *Living carbon is different from fossil-fuel carbon and both of these have essential uses.* Living carbon is replenishable over the short term by the processes of life, whereas fossilized carbon is not. Fossilized carbon should, therefore, be recycled as much as possible and restricted to essential uses. Anything that can be reasonably obtained from living carbon should be obtained in that manner. Today the fossil fuels are largely used as a source of energy, and this usage must be phased out as much as possible. Research must be directed towards finding good substitutes for the internal-combustion engine.

A dynamic programme of carbon-resource management must be set up to draw together these important features and effects of carbon resources and their uses. At any one point in time a dynamic equilibrium should exist in the human use of carbon products. Recycling must be carried out on both the living and fossilized sides of the cycle to maximize resource usage. (Here is a good reason why disintegratable plastics are a bad idea: they contribute to a one-way depletion of our petrochemical reserves and produce carbon dioxide.) The fossilized side of the cycle must, therefore, remain as closed as possible. On the living side, better means must be researched to integrate the used carbon back into the biosphere; for example, by turning them into a compost for building soils. As time goes on and more people reach a higher quality of life, and as research finds alternatives to thermal-electric generation and the internal-combustion engine, the equilibrium will shift.

These criteria must be applied in the construction of resource-management pictures for all materials. Only in this way will it be possible to account for all of the varying effects of resource extraction, consumption, and disposal in terms of the biogeochemical cycles and the patterns of life at any one time. These criteria must also be used in the development of a new system of constraints for the products of human industry. During the planning-stage of a product, its disposal pathways must be considered. Metals should be easily separable and re-usable at low-energy costs. Products (such as low-grade paper) that are to be composted or incinerated must be free of trace metals, persistent pesticides, PCB's, and other contaminants. Standardized containers, made of materials that are recycled with low energy costs, should be employed. Markings on containers should again be fairly easy to remove at the recycling stage and contain biodegradable materials. Similarly product compositions must be designed so as to minimize the usage of, as well as to maximize the availability of, scarce resources. These constraints are urgently needed, in addition to the ones that governments presently apply.

Also urgently needed today are public agencies to oversee these patterns of resource management.

(b) Closed Industrial Systems

These systems will apply to materials that are obtained from without the biosphere, one example of which is nickel. In the processing of nickel, nickel sulfide ores are smelted and atmospheric oxygen is used as an oxidizing agent. Sulfur dioxide and nickel are produced as well as a slew of minor metallic components, such as palladium and platinum, that comprise the smelting residues. All of the reactants and products, with the exception of oxygen, are from outside the biosphere. They must, therefore, be cycled along closed pathways apart from the biosphere, and if at any point they enter the biosphere, it must be in very limited quantities that do not upset the biogeochemical flows of these elements. Nickel is generally kept within the human world, except for the trace amounts from electroplating and metal processing, which, released into aquatic ecosystems, cause serious harm. Here pollution-abatement programmes are essential. Many of the trace metallic components of the smelter wastes are valuable trace metals. They are already worth recovering, and International Nickel in Sudbury nearly pays for its total operations with the profits from these rare metals alone. Sulfur dioxide, however, is another matter. It can be catalytically converted to sulfuric acid, an important industrial chemical. Although INCO in Sudbury attempts to recover some of the fumes in this manner, it is economically cheaper today to use raw sulfur, mined in Texas, for making sulfuric acid. It is only cheaper, however, because we fail to bring into the economic picture the long-term values of sulfur cycling. About Sudbury the damage from sulfur-dioxide pollution is extensive. (Elsewhere millions of dollars' worth of metal corrosion, plant damage, and respiratory afflictions have been caused by the failure of industrialists to recover sulfur dioxide.) Here then the government must play an important role, by placing tax-incentives on sulfuric-acid manufacture from industrial by-products, thus discouraging methods that are, in the long term, unsound. At the moment it is difficult to conceive of a return pathway for the atmospheric oxygen that is consumed in the smelting process, if it should be of consequence. Further research along these lines is necessary.

(c) Integrated Industry

Industries such as pulp-and-paper manufacture transform materials that are found predominantly within the biosphere. Here the by-products must either be channelled out along new pathways of industry, or fed back into living systems in patterns and ways that are stimulating. The processing wastes of pulp-and-paper manufacture include bark, wood chips, sawdust, wood fibres, pulped wood, cellulose-degradation products, and a wide variety of organic materials with high oxygen demands. There are already indications that many of these materials can be used as starting products for other industries, such as the manufacture of fumaric acid and vanilla. With adequate research, much of the pulp and wood wastes could be converted into a rich humus, which might be returned to forest soils, or used for agriculture, in combination with sewage wastes. The particularly toxic components of pulp-and-paper wastes have often been added during the manufacturing process—chlorine for bleaching, or organo-mercurial slimicides. Alternatives for these must also be found.

In summary, there are already indications of the directions in which industrial processing must begin to move. The pathways available offer great potential, but will require drive and imagination. Especially important is the vast amount of research and development that is necessary. In comparison with the gadget and gimmick trends of development in the sixties, these are the directions that both the public and government sectors should be demanding of industry during the next decades. There is a great deal we do not know, but it is clear that stop-gap solutions alone are not the answer to the one-way waste-production arrow. Often these non-solutions begin with one waste and end up with another disposal problem. Although in some cases recycling will not be applicable, simply because the amounts of wastes are so small that it will never be feasible to recover them, the move towards cyclic processing will prove to be a very satisfactory answer to many of our current problems.

Human Freedoms and Government

Human beings are biological organisms; they have both a dependence upon, and an obligation to, the biological world. Even human behaviour is at least partially a biological phenomenon, influenced by instincts and patterns that had their origins in evolution. Human behaviour may have been altered by cultural evolution and by the everyday human cultural environment. Beneath this, however, it is still biologically influenced by desires to compete and co-operate, by sexual desires and the hunger for food. The human way of life, through scientific and cultural exploration, is probably a justifiable one, and it is one that we, as a group of animals, may not desire to give up. Our so-called freedoms have very much been 'bought' by technology from the strictures of the ecological world. They have not yet been paid for, and only will be, to human satisfaction, when we shoulder full responsibility for them.

We must, therefore, redefine the concept of human freedom. All of us live on a finite planet. In the end every human action may bear important consequences upon the activities and well-being of other human and biological individuals. The most potent effects of human freedom are felt when people are born and when matter and energy are transformed in ways and patterns that are biologically atypical. We thus have a fundamental responsibility, because we employ technology, to fully understand our every action in terms of the biosphere as well as in terms of the human race. The pools of resources belong not to a few individuals of one generation, but to the whole history of our species. The biosphere, now warped by the mechanisms of technological humans, is in our trust.

Systems of government must reflect this new meaning of freedom. Should, for example, the Liberal Party now in power in Canada (one party for a four-year term of office) have the authority to make short-term and easy deals on matters such as resources and energy that are the property of not just this generation of Canadians but of the whole world for many centuries to come? By the same token, decisions regarding agricultural land, the growth of new cities, the cutting of forests, the practices of fishing, must all be made on a long-term basis. They are hardly matters of short-term policy and certainly not matters of individual human freedoms. They are decisions into which more than the immediate demands of society and far more than the demands of a single individual attempting to make a fast dollar must be weighed. New and stronger international agencies are needed to safeguard the commons of the world: the oceans, the atmosphere, the non-renewable resources, and the tracts of ecologically intact life. Yet these organizations must be constituted so that no one group of people gains too much power and so that the rights of individual nations and peoples to a just hearing are guaranteed. Much of the card-playing diplomacy of today's world reflects the incredible difficulty of dealing with matters of national interest, and it is naïve to think that international government alone would solve all of the problems. International control measures are necessary, and new forms of international government must arise in order to deal with them. They will, by necessity, have to be complex, in order that all of the players be counted. Their conception however must begin today.

CHAPTER 3—2

Checklist for Survival

In order to formulate long-term solutions to the global environmental crisis and to achieve human survival, meaningful and effective international action must be taken. This action must cover a broad range of human affairs, and through it individual nations must be willing to co-operate fully on an international level. Against a common background of crisis and through a large degree of collaboration in the solving of environmental problems, we may be able to move towards an era of unprecedented unity that has always been desired. The following are some very basic actions that must be taken, and the dates by which they must be instituted. Please signify, as shown, if action or results seem likely by the given date YES ✓
NO ×

1. Population growth must be recognized as a fundamental factor in the environmental crisis. Subsequently, the means to achieve zero population growth must be instituted. This must be accomplished in developed nations such as Canada, the United States, Europe, and the Soviet Union by 1975. ☐

2. The means to achieve Z.P.G. must be undertaken by the developing nations, including China. This goal will require a longer time period than in the developed nations for several reasons, among them lack of communications systems and greater numbers of people to be reached. New research into cheap and practical contraception is needed and must be initiated on a large scale. Means for reaching Z.P.G. must be instituted within a majority of the developing nations by 1980. ☐

3. A broad and international program of environmental education is essential to alert every human citizen to the crisis we face. This program must reorient thinking to life on spaceship Earth. It should make use of modern satellite communications systems in order to have its broadest impact, and could logically arise from the United Nations Conference on the Human

Environment planned for Stockholm in 1972. The program should be initiated by 1973. ☐

4. A population optimum for Earth must be determined. This should be carried out by a United Nations committee representative of all nations and peoples and considerate of all aspects of human needs. This report should lay the groundwork for future international trends of population growth (increasing or decreasing) and should formulate the means necessary to reach optimum. A logical outgrowth of the planned United Nations Population Conference of 1974, the world optimum and future trends should be sketched out by 1980. ☐

5. A far-sighted reorientation of the World Health Organization and the Food and Agriculture Organization (both arms of the United Nations) is necessary. Research must be directed into better modes of agriculture and fishing so that animal and plant stocks are maintained and ecological stability is maximized at the same time as yield. Intensive research into the recycling of sewage and the dynamics of the phosphate cycle must also be instituted. These programs should be underway by 1975. ☐

A new international program of research into the integrated and biological control of crop, forest, and human pests must be started by 1973. ☐

6. A World Soils Bank must be set up to inventory the world's soils and to investigate their state of being. It must attempt to measure the annual losses due to erosion and to carry out research into methods of agriculture that maintain soil fertility. This body should be given international powers to regulate the use and preservation of soils. This body should be at work by 1980. ☐

7. A World Resource Inventory must be set up and carried out to consider the ultimate availability of minerals, energy, and renewable resources on Earth and to work hand in hand with the optimum population committee, all by 1975. ☐

8. An International Bureau of Resource and Energy Manage-

ment to institute worldwide control on the extraction and consumption of energy and resources, including forests, on an ecological basis should be set up by 1980. ☐

The bureau should be operating by 1985. ☐

It should work out patterns of world recycling by 1983. ☐

This bureau should be given the essential task of research into new energy sources such as the breeder and fusion reactors. It should supervise the shift of energy supply to environmentally sounder methods. On an international scale this body should monitor the construction, operation, and safety of nuclear-power facilities.

9. An International Commission on Economic Growth should be set up within the developed nations to reassess the priorities of continued exponential growth in the light of (7) and (8) and to formulate future trends in terms of steady-state economics. The commission should be underway by 1975. ☐

A cut-back in the Western consumption of resources and energy should be started by 1980. ☐

Good systems of industrial and municipal recycling should have been instituted or should be in the planning stages by 1975. ☐

10. An International Commission on Economic Development must be devised to study patterns of development (in presently developing nations) in terms of available resources and energy. This commission should formulate policy in terms of the studies of groups (7) and (8). In particular it must search for new modes of economic development that avoid the pitfalls of Western economic methods. This commission should be set up to 1975. ☐

It should have new programs for development by 1985. ☐

11. An Urban Planning Bureau to investigate trends in rural-to-urban migration and the most satisfactory ways of accommodating the inevitable population growth between now and the year 2000. This bureau should investigate transportation

such as commuter transit, the S.S.T., the monorail, and the internal-combustion engine, and should reach consensus on international controls and directions in transportation. This group should be initiated no later than 1975. ☐

12. An International Chemical Control Agency must be set up to have jurisdiction in all nations over the production and release of any presently used or new chemicals on a commercial scale. This agency will test and regulate all new drugs, pesticides, herbicides, paints, synthetic materials, and food additives from an environmental viewpoint. This group to be in action by 1975. ☐

13. An International Joint Commission on the Atmosphere and the Oceans should be set up to carry out primary research of the commons that are shared by all of humankind. They must be given policing powers over marine and atmospheric pollution and overfishing. This body should regulate the shipment of oil by sea and the drilling for oil in coastal waters. It should investigate oil-spills and provide clean-up teams and expertise in the case of spills. This body should establish marine protectorates for the preservation of diverse and essential ocean ecosystems and should carry out research on the crown of thorns starfish. This body must be in working order by 1975. ☐

14. A World Biological Bank that could be administered by UNESCO should be set up to designate and maintain ecological preserves in all nations. In a biosphere where we are rapidly losing the last ecologically sound systems, these preserves are of international, not national, concern. They must, therefore, be under strict international supervision even though they may lie within national boundaries. Furthermore, the bank should be responsible for the preservation of genetic pools of living organisms and the protection of endangered species. This important body must be underway by 1975. ☐

15. In conjunction with (14) there must be a massive step-up of the International Biological Program, which is presently carrying out fundamental ecological research, by 1973. ☐

16. Within many nations we hope to see a shift of international and national policies along environmental lines. These

new outlooks should recognize the fundamental limits of living within the biosphere and the limited supply of resources and energy. In nations such as Canada where national feelings are, to some degree, already in sympathy with these limits, this policy should emerge by 1975. Should nations like Canada adopt a national policy of this sort, it may play a key role in international understanding of the environmental crisis. □

All sixteen of these steps are very large ones and in the light of the present international situation it is very questionable that we shall be able to take them. But in some form or other they are all necessary steps. Look carefully at this list then, and watch as the next decade goes by. If 1980 arrives and half of the goals are unfulfilled and have yet to reach international consideration, then the world is in bad shape. Should they be at least half-filled, then there is definitely hope. Should they all be achieved then the future holds very great promise. Never before in so short a period and when the odds are so great against us, have we had to achieve so much in order to ensure survival.

Bibliography

I

1. Eshleman, Von R. 'The Atmosphere of Venus and Mars.' *Scientific American*, March 1969. New York.
2. Cloud, P. E., and Gibor, A. 'The Oxygen Cycle.' *Scientific American*, September 1970. New York.

   Ehrensvard, Gosta. *Life : Origin and Development*. Chicago: University of Chicago Press (Phoenix Books), 1962.

   Hanawalt, Philip C., and Haynes, Robert H., eds. *The Molecular Basis of Life, An Introduction to Molecular Biology*, readings from *Scientific American*. San Francisco: W. H. Freeman, 1968.

   Jeffers, Robinson. *Not Man Apart*. New York: Sierra Club/ Ballantine, 1969.

   Keosian, John. *The Origin of Life*. Newark: Reinhold Book Corporation, 1968.

   Lehninger, A. L. *Bioenergetics*. New York: W. A. Benjamin Inc., 1965.

   Marquand, J. *Life: Its Nature, Origins and Distribution*. Edinborough: Oliver and Boyd, 1968.

   Oparin, A. I. *Origin of Life*. New York: Academic Press, 1957.

   Watson, J. D. *The Molecular Biology of The Gene*. New York: W. A. Benjamin Inc., 1965.
3. Anfinsen, C. B. *The Molecular Basis of Evolution*. New York: John Wiley and Son, 1963.

   Darwin, Charles. *The Origin of Species* (rev. ed.). New York: Collier-Macmillan, 1962.

   Kormandy, E. J., ed. *Readings in Ecology*. Princeton, N.J.; Prentice-Hall, 1965.

   Laporte, Leo F. *Ancient Environments*. Princeton, N.J.: Prentice-Hall, 1968.

   Lerner, I. Michael. *Heredity, Evolution and Society*. San Francisco: W. H. Freeman, 1968.

   Odum, E. P. *Fundamentals of Ecology*. Philadelphia: W. B. Saunders and Co., 1971.

   Simpson, George Gaylord. *The Major Features of Evolution*. New York: Simon and Schuster, 1953.

4. Lehninger, op. cit.

MacIntyre, F. 'Why the Sea is Salt.' *Scientific American*, November 1970, New York.

Redfield, A. C. 'The Biological Control of the Chemical Factors of the Environment' in *Readings in Ecology*, E. J. Kormandy (ed.). Princeton, N.J.: Prentice-Hall, 1965.

Woodwell, G. M. 'The Energy Cycle of the Biosphere.' *Scientific American*, September 1970, New York.

5. Cox, G. W., ed. *Readings in Conservation Ecology*. New York: Appleton-Century-Crofts, 1969.

Kormandy, E. J. *Concepts of Ecology*. Princeton, N.J.: Prentice-Hall, 1969.

Leopold, Aldo. *A Sand County Almanac*. New York: Sierra Club/Ballantine, 1970.

Odum, op. cit.

II

1. Bates, Marston. *Man in Nature*. Princeton, N.J.: Prentice-Hall, 1965.

Braidwood, Robert J. 'The Agricultural Revolution.' *Scientific American*, September 1960, New York.

Ehrlich, Anne H., and Paul R. *Population/Resources/Environment*, San Francisco: W. H. Freeman, 1970.

Langer, William L. 'The Black Death.' *Scientific American*, February 1964, New York.

McKinley, Daniel, and Shepard, Paul, eds. *The Subversive Science: Essays toward an Ecology of Man*. Boston: Houghton Mifflin Co., 1969.

2. Cloud, P. E., ed. *Resources and Man*. San Francisco: W. H. Freeman, 1969.

Dominion Bureaux of Statistics. *Population, Specified Age Groups and Sex: Census of Canada 1966*. Ottawa: The Queen's Printer, August 1968.

Ehrlich, Paul R. *The Population Bomb*. New York: Sierra Club/Ballantine, 1968.

Ehrlich and Ehrlich, op. cit.

Falls, J. B., and Regier, H. A., eds. *Exploding Humanity: The Crisis of Numbers*. Toronto: House of Anansi, 1969.

Hardin, Garret, ed. *Population, Evolution and Birth Control*. San Francisco: W. H. Freeman, 1965.

Jacobs, Jane. *The Death and Life of Great American Cities*. New York: Random House, 1965.

Lerner, op. cit.

McKinley and Shepard, op. cit.

Nowlan, David, and Nowlan, Nadine. *The Bad Trip*. Toronto: New Press, 1969.

Canada: Population Projections to the Year 2000. Toronto: Systems Research Group, 1970.

3. Borgstrom, Georg. *The Hungry Planet*. New York: Collier-Macmillan, 1967.

Brown, Lester R. 'Human Food Production as a Process in the Biosphere.' *Scientific American*, September 1970. New York.

Cloud, op. cit.

Cox, op. cit.

Ehrlich and Ehrlich, op. cit.

Gustafson, A. F. *Soils and Soil Management*. New York: McGraw-Hill, 1941.

Leopold, op. cit.

Litteljohn, Bruce M., and Pimlott, Douglas H., eds. *Why Wilderness?* Toronto: New Press, 1970.

McKinley and Shepard, op. cit.

McNeill, Mary. 'Laterite Soils.' *Scientific American*, November 1964. New York.

Paddock, Paul, and Paddock, William. *Famine 1975!* Boston: Little, Brown & Co., 1967.

Pinchot, Gifford B. 'Marine Farming.' *Scientific American*, December 1970. New York.

Schaller, Friedrich. *Soil Animals*. Ann Arbor: University of Michigan Press, 1968.

Whiteside, T. *Defoliation*. New York: Sierra Club/Ballantine, 1970.

4. Bonner, James; Brown, Harrison; and Weir, John. *The Next Hundred Years*. New York: Viking Press, 1963.

Brown, Harrison. 'Human Materials Production as a Process in the Human Biosphere.' *Scientific American*, September 1970. New York.

Cloud, op. cit.

Ehrlich and Ehrlich, op. cit.

Landsberg, Hans H. *Resources for American Growth to the Year 2000*. Baltimore, Md: Johns Hopkins Press, 1964.

5. Brown, Tom. *Oil on Ice*. New York: Sierra Club Battlebook, 1971.
Cloud, op. cit.
Curtis, R., and Hogan, E. *The Perils of the Peaceful Atom*. New York: Sierra Club/Ballantine, 1970.
Ehrlich and Ehrlich, op. cit.
Fabricant, Neil, and Hallman, Robert M. *Toward a Rational Power Policy*. New York: George Braziller, 1971.
Gofman, John W., and Tamplin, Arthur R. *Poisoned Power*. New York: Rodale Press, 1971.
Hogerton, J. F. 'Thermal Pollution.' *Scientific American*, February 1968. New York.
Laxer, James. *The Energy Poker Game*. Toronto: New Press, 1970.
National Energy Board of Canada. *Energy Supply and Demand in Canada and Export Demand for Canadian Energy, 1966-1990*. Ottawa: The Queen's Printer, 1969.
Odum, Howard T. *Environment, Power and Society*. New York: Wiley-Interscience, 1971.
Report of the Study of Critical Environmental Problems. *Man's Impact on the Global Environment*. Cambridge, Mass.: M.I.T. Press, 1971.
Seaborg, Glenn. 'The Breeder Reactor.' *Scientific American*, November 1970. New York.
Scientific American, general eds. of 'Energy'.
Scientific American, September 1971. New York.

7. Borgstrom, op. cit.
Cloud, op. cit.
Holt, S. J. 'The Food Resources of the Ocean.' *Scientific American*, September 1969. New York.
Johnson, C. D. 'The Crown of Thorns.' *National Geographic*, March 1970. Washington, D.C.

8. Borgstrom, op. cit.
Cloud, op. cit.
Ehrlich and Ehrlich, op. cit.
Forrester, Jay W. *World Dynamics*. Cambridge, Mass.: Wright-Allen Press, 1971.
Paddock and Paddock, op cit.
Report of the Workshop of Global Ecological Problems. *Man in the Living Environment*. Institute of Ecology, 1971.

III

1. Brinkhurst, Ralph O., and Chant, D. A. *Our Good Good Earth.* Toronto: Macmillan of Canada, 1971.

Chant, D. A., ed. *Pollution Probe.* Toronto: New Press, 1970.

Cloud, op. cit.

Ehrlich and Ehrlich, op. cit.

Leopold, op. cit.

McKinley and Shepard, op. cit.

Acknowledgements

In writing this book I have many friends to thank, not only for helpful and freely given advice, but also for their stimulating discussions, patience, and encouragement. Especially I have to thank my colleagues both in the Department of Biochemistry, University of Toronto, and in Pollution Probe.

In particular there are many individuals who have read manuscript for me, both generally and in their fields of expertise. To these people I am indebted. I would therefore like to thank Dr. Ralph Brinkhurst, Dr. Donald Chant, Mrs. Maureen Gould, Dr. Theodor Hofmann, Mr. Lamont Hummel, Mrs. Ruth Kelly, Mr. David Lean, Mr. Peter Middleton, Dr. Christopher Plowright, Dr. Henry Regier, Mr. Robert Stasko, Mr. Paul Tomlinson, and Dr. Lynn Trainor.

It is a great pleasure to acknowledge the devotion and time with which Mr. Brian Kelly of Pollution Probe critically read and edited the complete manuscript and provided cogent suggestions. Great appreciation is expressed also to my mother Mrs. Dorothy Mains for the long hours and hard work put into typing the manuscript. Thanks must also go to Jim Bacque of New Press for his advice and editorial skill.

Index